Andreas Valet
Adalbert Braig

Light Stabilizers for Coatings

2nd Revised Edition

VINCENTZ
Providing Professional Information

Cover: Coloures-pic/Fotolia

Andreas Valet, Adalbert Braig
Light Stabilizers for Coatings, 2nd Revised Edition
Hanover: Vincentz Network, 2017
European Coatings Library
ISBN 978-3-86630-128-3

Vincentz Network, Plathnerstr. 4c, 30175 Hanover, Germany
T +49 511 9910-033, F +49 511 9910-029, books@european-coatings.com

Layout: Danielsen Mediendesign, Hanover, Germany

European Coatings Library

Andreas Valet
Adalbert Braig

Light Stabilizers for Coatings

2nd Revised Edition

Foreword

Ah, happy he who still can hope to rise,
Emerging from this sea of fear and doubt!
What no man knows, alone could make us wise;
And what we know, we well could do without.
Goethe, Faust

Almost twenty years after the first edition of this book, the editor was asked by colleagues in the coating industry whether it would be possible to publish an up-dated second edition. When the editor asked us whether we would be interested in working on such a second edition, we agreed with great pleasure.

Although some time has passed since the first edition, the problems facing the industry are still the same: paint flaking off an object remains a serious concern. The object has lost its protection and is exposed to the elements. It has lost the colour that made it attractive. It has become insignificant and unnoticed.

The purpose of this book is to explain the underlying principles of paint degradation, demonstrate, with the help of numerous examples, how paint films can be protected and serve as a practical guide for formulators when selecting light stabilizers for their paint formulation.

For a more precise analysis of the mechanism of paint degradation and its prevention, the reader is referred to the extensive bibliography at the end of the book, which covers the subject comprehensively.

Hermann Hesse wrote "Blue, yellow, white, red and green – what wonderful colours"[1]. When properly stabilized, colourful finishes ARE wonderful.

We would like to take this opportunity to thank our colleagues from the former Ciba Specialty Inc. and BASF Switzerland AG. Without their collaboration and support it would have never been possible to carry out the research, over a period of more than two decades, on which this book is based. We would also to thank Dr. Godwin Berner and Hans-Jürgen Berger who pioneered, built-up and led this successful business for so many years. Special thanks goes to Allan Cunningham for his great support editing our English text.

Basle, Switzerland June 2016

Andreas Valet and Adalbert Braig

Contents

1 **Introduction** 11

2 **Light and photo-oxidative degradation** 13
2.1 Light 13
2.1.1 Photo-physical processes 15
2.2 Photo-chemical degradation processes 16

3 **Stabilization options** 21
3.1 UV absorbent pigments 22
3.2 UV absorbers 23
3.2.1 UV absorber classes 24
3.2.2 Mode of action of UV absorbers 27
3.2.2.1 Phenolic UV absorbers 28
3.2.2.2 Non-phenolic UV absorbers 30
3.2.3 Examples of UV absorbers 30
3.3 Free-radical scavengers 36
3.3.1 Antioxidants 36
3.3.2 Sterically hindered amines 38
3.3.2.1 Mode of action of HALS 42
3.4 Quenchers 45
3.5 Peroxide decomposing agents 46

4 **Stabilization of coatings** 49
4.1 Automotive coatings 51
4.2 Light stabilization of automotive coatings 52
4.2.1 Two-coat systems 53
4.2.2 Specific requirements of UV absorbers in coatings 55
4.2.2.1 Solubility and compatibility of UV absorbers 56
4.2.2.2 Volatility of UV absorbers 59
4.2.2.3 Reactable UV absorbers 61

4.2.2.4 Effect of UV absorbers on coating colour 63
4.2.2.5 Unwelcome side reactions 64
4.2.2.6 UV absorbers and photoinitiators 67
4.2.3 Specific requirements on HALS in coatings 71
4.2.3.1 Solubility and compatibility of HALS 71
4.2.3.2 Volatility of HALS 74
4.2.3.3 Reactable HALS 75
4.2.3.4 Effect of HALS on coating colour 76
4.2.3.5 Unwelcome side reactions 77
4.2.4 Weathering results for two-coat systems 79
4.2.4.1 Weathering tests 79
4.2.4.2 Results for solvent-borne clear coats 86
4.2.4.3 Results for water-borne clear coats 97
4.2.4.4 Results for powder clear coats 100
4.2.4.5 Results for UV-curable clear coats 103
4.2.4.6 Coatings on plastic substrates 106
4.2.4.7 UV protection of epoxy-based fibre reinforced plastics 111
4.2.4.8 Effect of additional basecoat stabilization 114
4.2.4.9 Exposure results for one-coat finishes 117
4.3 Light stabilization of industrial coatings 121
4.3.1 Stabilization of paints for metal substrates 122
4.3.2 Stabilization of clear wood coatings 125
4.4 Stability of light stabilizers 128
4.4.1 Photo-chemical stability of UV absorbers 128
4.4.2 Long-term stability of HALS 137

5 Conclusions 141

6 References 143

Authors 151

Index 153

1 Introduction

In the dictionary, paints are defined as "liquid or powdered, solid substances which are applied thinly to objects and which dry by chemical reaction and/or physical changes to form a solid film whose function may be decorative and/or protective"[2].

Although applied only thinly, paints alter the appearance and increase the durability of many everyday objects. The efficiency of a paint, i.e. its ability to protect the coated object, is governed by the nature of the binder used in its formulation. Binders are also often referred to as film-forming agents, surface coating resins or synthetic resins. Paints contain organic solvents and/or water, or are completely solvent-free, depending on the kind of binder used. Paints may also contain pigments, fillers and other additives.

Paint films are exposed to all conditions arising in daily life, including mechanical stresses, chemicals and weathering, against which they must protect the coated object.

In 1860, *A. Hofmann* stated[3] that "the characteristic change which occurs in gutta-percha (rigid natural latex) when it has been in contact with air for some time, is well known. It becomes brittle and irreversibly loses its texture." Investigations had shown this change to be due to oxidation of the gutta-percha when exposed to air and Hofmann's statement is probably the first reference in the literature concerning the chemical reactions that alter polymer properties.

As a result, efforts began to protect polymers against these chemical reactions. New polymers required stabilizers to prevent them from deterioration in practical use. This formed the basis for the development of stabilizers for polymers. The term "stabilizer" refers to any additive that prevents or delays polymer degradation, irrespective of what kind of degradation mechanism is involved. The development of light stabilizers is described in the following publications and patent specifications.

- Remarks on the Change of Gutta Percha under Tropical Influences[3]
- Verfahren, um das Erhärten und Brüchigwerden von Kautschuk, Guttapercha, Balata und ähnlichen Gummiarten zu verhindern (Methods of preventing the hardening and embrittlement of rubber, gutta-percha, balata and similar rubbers) DP 221310; W. Ostwald, 1908)
- The Chain Reaction Theory of Negative Catalysis[4]
- Autoxidation von Kohlenwasserstoffen: Über ein durch Autoxidation erhaltenes Tetrahydronaphthalin-Peroxid (Autoxidation of hydrocarbons: tetrahydronaphthalene peroxide obtained through autoxidation)[5]

- Der Kettenmechanismus bei der Autoxidation von Natriumsulfidlösungen (The chain mechanism during the autoxidation of sodium sulphide solutions) [6]
- Pellicle and the Manufacture thereof (USP 2,129,131; E. Du Pont de Nemours, 1938
- Vinylidene Chloride Composition Stable to Light (USP 2,264,291; The Dow Chemical Company, 1941)
- Lichtschutzmittel und ihre Beurteilung (Light Stabilizers and their Assessment) [7]
- Weather resistance of Cellulose Ester Plastics Compositions [8]
- 4-Benzoylresorcinol as an Ultraviolet Absorbent (USP 2,568,894; General Aniline & Film Corporation, 1951)
- Verwendung von 2-Phenylbenzotriazol-Verbindungen zum Schützen von organischem Material gegen ultraviolette Strahlung (Use of 2-phenylbenzotriazole compounds to protect organic materials from UV radiation) (DE 1185610; Ciba-Geigy AG, 1957)
- Free Radical Reactions involving no Unpaired Electrons [9]
- Stabilization of Synthetic Polymers (USP 3,542,729; Sankyo Ltd., 1970)

The preceding references formed the basis for the development of light stabilizers for coatings.

In this book, in addition to paint, two other terms are widely used in the industry will appear, namely coating and varnish.

2 Light and photo-oxidative degradation

2.1 Light

Light is generally described as radiation visible to the human eye comprising wavelengths between 400 and 750 nm [2]. But "visible" light is only a part of the electromagnetic radiation to which the earth is exposed. Electromagnetic radiation can be divided into different groups as shown in Figure 2.1. Figure 2.2 shows the sub-division of "ultraviolet to infrared".

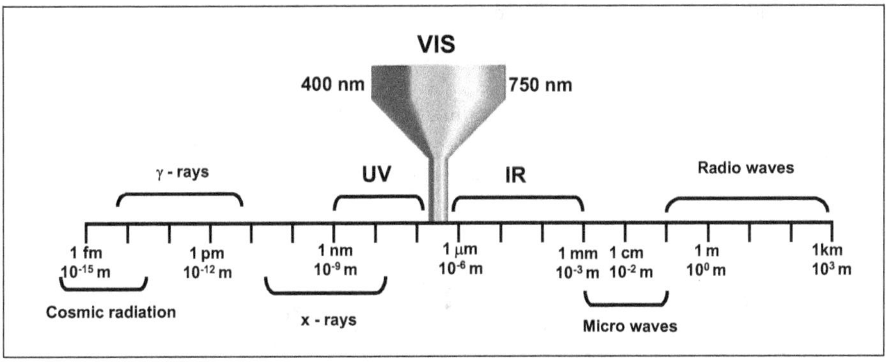

Figure 2.1: Classification of electromagnetic radiation with wavelengths λ of 10^{-15} to 10^3 m [10]

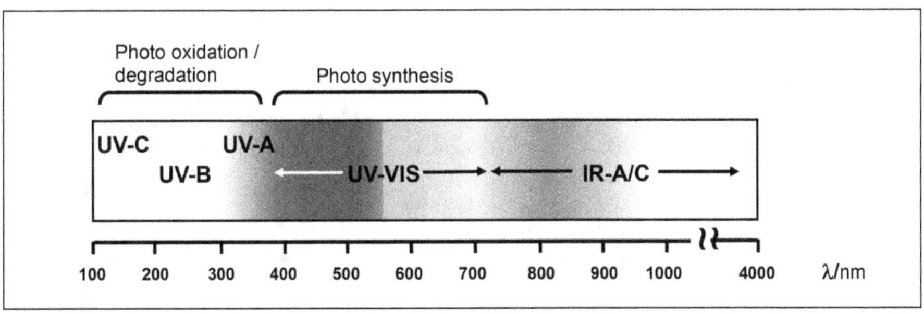

Figure 2.2: Classification of radiation with wavelength range of 100 to 4000 nm

Table 2.1: Relationship between wavelength λ and dissociation energy of some organic model compounds [11, 12, 13]

	λ [nm]	**Bond**	**Type of bond**	**Dissociation energy [kJ/mol]**
UV-B	230	-C-**C**-	aromatic	520
	286	R-O-**H**	alcohol	420
	290	R-CR_2-**H**	prim./sec./tert. H	410/395/385
	310	C-**O-H**	alcohol	385
	320	-C-**O**-	ether	365 to 390
UV-A	340	R-CH_2-**CH_3**	aliphatic	335 to 370
	350	-CR_2-**Cl**	aliphatic chlorides	330 to 350
	360	-CH_2-**NR_2**	amine	330
	400	**-O-O-**	peroxide	270

Most of the energy-rich electromagnetic radiation ($\lambda < 290$ nm) is absorbed by the earth's atmosphere, primarily by the ozone layer in the stratosphere. Although only 6 % of the light reaching the earth's surface is the ultraviolet light ("UV light") it is responsible for most of polymer degradation due to weathering.

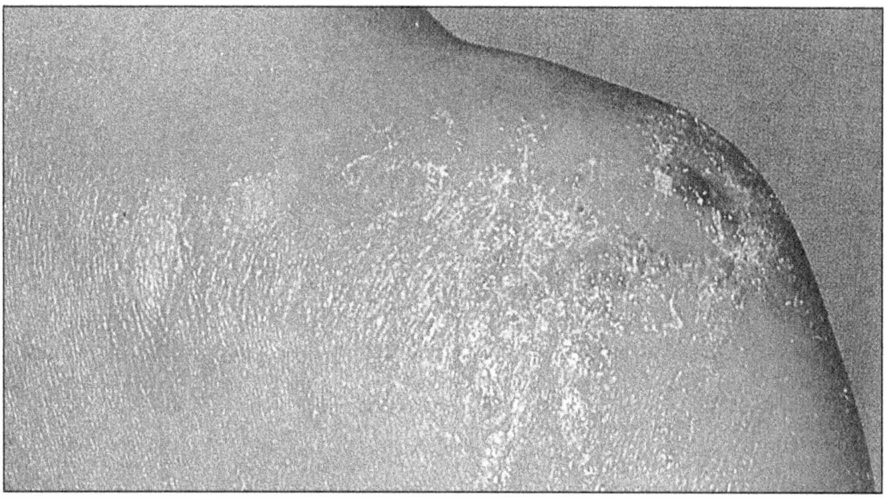

Figure 2.3: Sunburn of the human skin caused by UV radiation (UV-A, UV-B) [14]

Table 2.1 shows the relationship between wavelength and dissociation energy in different molecules [11, 12, 13]. These figures demonstrate that the UV light reaching the earth's surface is sufficiently rich in energy to break covalent bonds present in polymers.

Only light that is absorbed is capable of initiating photo-chemical processes. **Pure** polymers such as polymethyl methacrylate, polyethylene or aliphatic polyesters are photochemically stable between 300 and 400 nm because they do not absorb light. The emphasis here is on the word *"pure"*. However, if the polymer contains impurities that absorb light, e.g. catalyst residues and other substances added during production, or oxidation products, it becomes sensitive to UV light. Polymers whose basic structure already contains UV absorbent groups are inherently light sensitive and are likely to undergo photo-chemical degradation even in the absence of such impurities. Typical examples include polystyrene and styrene copolymers, aromatic polyurethanes, polyesters and polyepoxides.

Human beings, too, experience the harmful – and very painful – effects of light. Here, the skin, whose function, after all, is to protect the body, is attacked by UV-B, UV-A, and even visible light [14, 15]. Damage and eventual destruction of the skin can seriously affect the underlying tissue.

Figure 2.3 shows the sunburn of the human skin. Reflected light, water, snow and droplets of perspiration, for example, can further magnify the effect of sunlight.

By analogy, the damage of polymers and paint films by light can also be classified as a kind of "sunburn", the difference being that polymers, unlike the human skin, are not capable of regeneration.

2.1.1 Photo-physical processes

The first step in a photo-chemical reaction is the absorption of light that promotes the molecule into an energy-rich, excited state. Molecules can exist in two electronic states:

- singlet state S (paired electron spins)
- triplet state T (unpaired electron spins)

According to Hund's first law, electronic states with greater spin multiplicity are more stable, i.e. the triplet state is generally lower in energy than the corresponding singlet state.

Upon absorption of light, molecules are promoted from their singlet ground state S_o to a first energy-rich excited state S_1 or T_1.

The probability of a chemical reaction in an excited state increases with the lifetime of the state. As the lifetime of the excited triplet state T_1 is longer than that of the corresponding singlet state S_1, most photo-chemical reactions occur in the excited triplet state [16]. Reactions in the shorter-lived singlet state S_1 can occur if they are thermodynamically and kinetically favoured. The kinetic factor in particular is dependent upon the substrate.

A molecule in the excited singlet state S_1 has more possibilities to dissipate energy. These are shown in Figure 2.4 in the so-called Jablonski diagram [17].

Energy release or deactivation is possible with or without radiation. The lowest triplet state T_1 is formed by a radiation-less transition from S_1 to T_1, described as "inter-system crossing". The transition T_1 to T_2, T_3 etc. is only possible if the molecule absorbs light in the T_1 state. Transitions from S_1 to T_1 or from T_1 to S_0 break the spin selection rules (change of spin multiplicity) but can occur if sufficient spin-orbit coupling is present [18].

The smaller the energy difference between S_1 and T_1, the greater the possibility of inter-system crossing.

2.2 Photo-chemical degradation processes

Photo-oxidative degradation processes, in which chain cleavage, chain branching and oxidation reactions occur, can be divided into the stages shown in Figure 2.5 [19]. Chain initiation, equation (a), involves energy transfer from a photo-activated donor D^* to an acceptor A present in the ground state. This energy transfer can take place inter-molecularly or intra-molecularly [12].

In intra-molecular energy transfer, a functionality in an excited state (D^*) transfers energy to a functionality in the ground state (A) within the same polymer molecule. This process is important in those molecules (e.g. copolymers) in which parts are present which can easily be photo-activated.

In the case of inter-molecular energy transfer, a molecule not belonging to the polymer itself but being in an excited state, transfers its energy to the polymer [19]. In Section 2.1 it was pointed out that **pure** polymers such as polyethylene and polymethyl methacrylate are photo-chemically stable. However, impurities present in these polymers, e.g. catalyst residues, can initiate photo-chemical degradation.

Chain initiation (a) in Figure 2.5 leads to free radicals on the polymer chain that react with oxygen (photo-oxidation). The speed of this photo-oxidation depends on the type of polymer and coating (preparation, composition, impurities, additives). At the same time, purely thermal processes may also take place (oxidation).

Given the enormous quantities of paints and polymers used throughout the world, stabilization against photo-chemical degradation is crucial from an economic as well as ecological point of view. After all, the destruction of paint films and polymers requires their replacement, especially when the coating serves not only for decorative reasons but to protect the underlying surface, be it metal, wood or plastic. Once the protective coating deteriorates and ultimately breaks down, the revealed surface is exposed to the elements: metal will corrode and wood will be degraded by humidity and UV light.

Figures 2.6 to 2.9 show typical examples of photo-oxidatively damaged polymers and coatings.

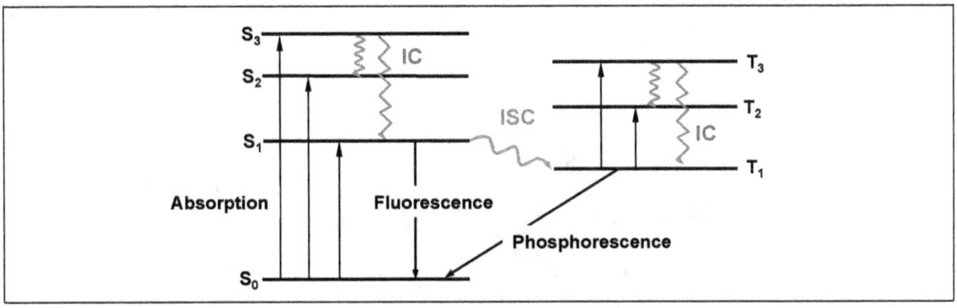

Figure 2.4: Jablonski diagram [17], IC: Internal Conversion; ISC: Inter System Crossing

Chain initiation:

$\text{polymer}^{1)} \xrightarrow{\text{UV light}} \text{free radicals } (P\cdot, PO\cdot, HO\cdot, HO_2\cdot, \dots)$ (a)

Chain propagation:

$P\cdot + O_2 \longrightarrow POO\cdot$ (b)

$POO\cdot + PH \longrightarrow POOH + P\cdot$ (c)

Chain branching:

$POOH \xrightarrow{h \cdot \nu \text{ or } \Delta T} PO\cdot + OH$ (d)

$2\,POOH \xrightarrow{h \cdot \nu \text{ or } \Delta T} PO\cdot + POO\cdot + H_2O$ (e)

$PO\cdot + PH \longrightarrow POH + P\cdot$ (f)

$HO\cdot + PH \longrightarrow H_2O + P$ (g)

Chain termination:

$P\cdot + P\cdot \longrightarrow P\text{-}P$ (h)

$P\cdot + POO\cdot \longrightarrow POOP$ (i)

$P\cdot + PO\cdot \longrightarrow POP$ (k)

$PO\cdot + PO\cdot \longrightarrow POOP$ (l)

Figure 2.5: Schematic representation of photo-oxidation of polymers (P)

1 may contain catalyst residues, hydro peroxides etc.

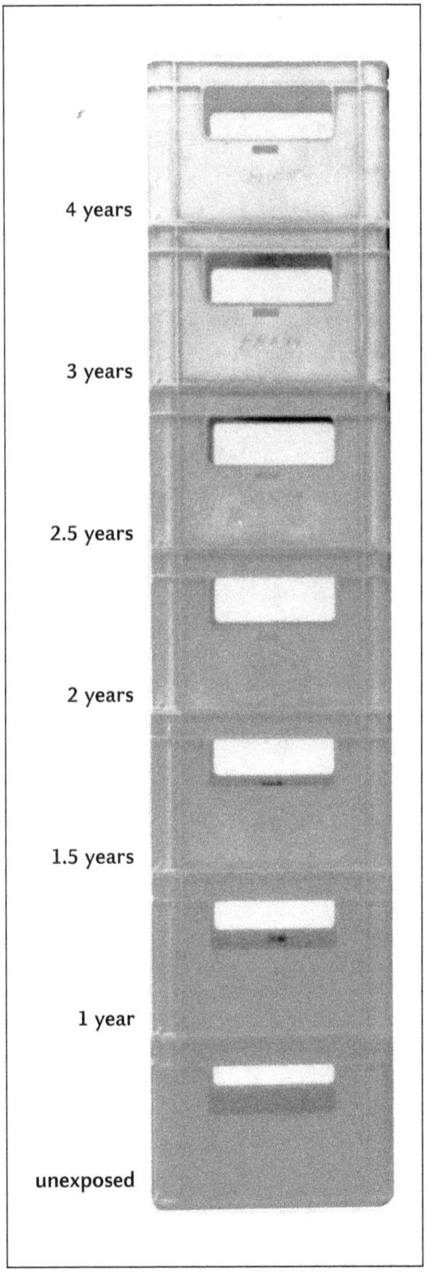

Figure 2.6 (left above):
Embrittlement of polymer material caused the destruction of the greenhouse film

Figure 2.7 (left below):
Loss of adhesion of a 2-coat automotive finish caused by weathering

Figure 2.8 (above):
Micrograph of cracking in an automotive clear coat

Figure 2.9 (right):
Outdoor weathering tests with red HD polyethylene. Weathering: Florida, 45° S

3 Stabilization options

The photo-physical and photo-chemical processes leading to photo-oxidative degradation of polymers described in Section 2.2, also provide an indication of the best methods to protect or stabilize these materials against the harmful effects of light. These are presented in Figure 3.1.

As afore mentioned, photo-oxidative reactions can only occur if the polymer absorbs light. The polymer must contain functionalities capable of absorbing electromagnetic radiation that leads to intra-molecular electron transfers. Such functionalities are termed **chromophores**. The absorption of light by a chromophore Ch (e.g. thermally damaged binders, catalyst residues etc.) promotes the chromophore to an excited state Ch*. Essentially, Ch* has four potential fates:

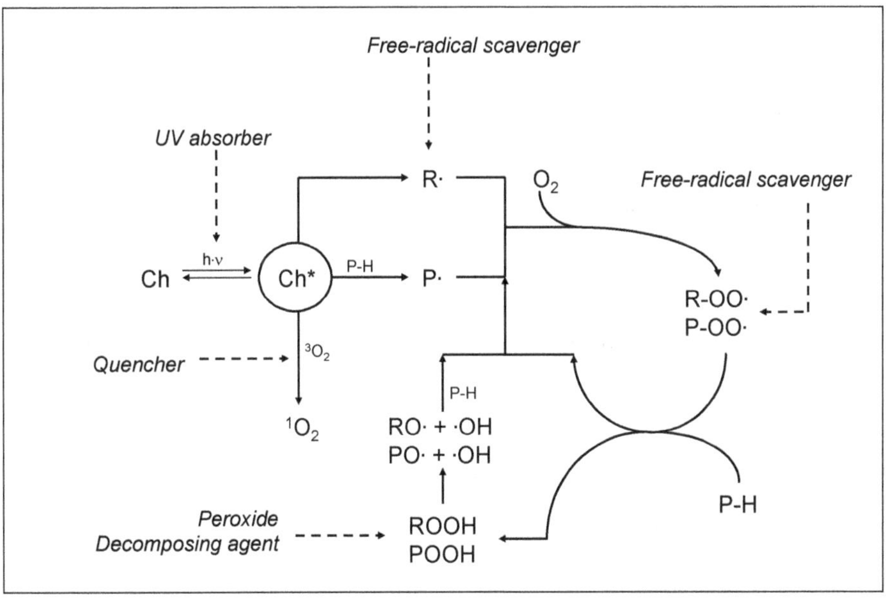

Figure 3.1: Schematic representation of photo-oxidative polymer degradation and methods of protection [20, 21]
Ch = chromophore; P = polymer; -> method of preventing photo-oxidation

a. return to the ground state via fluorescence (see Figure 2.4) or radiation-less deactivation (heat)
b. decomposition into radicals, followed by subsequent reaction with the substrate (polymer) and/or oxygen
c. formation of radicals through abstraction of hydrogen atoms from the substrate
d. energy transfer (e.g.to oxygen, which leads to the formation of singlet oxygen $^1O_2$)

Option a) does not endanger the polymer wheras processses b) to d) will result in damage to the polymer or a significant alteration of its properties.

Options a) to d), however, indicate possibilities to prevent or deviate the outcome of these processes [21–23]:

a. filter out the harmful UV light and, thereby, prevent the formation of Ch^* *(UV absorber)*
b. deactivate the exited state Ch^* by use of a suitable acceptor *(quencher)*
c. trap any radicals before subsequent reactions leading to polymer damage can occur *(free-radical scavenger)*
d. decompose any peroxides formed by use of suitable *peroxide decomposing agents.*

Additives which meet these functions are generally referred to as **light stabilizers**. This general term, however, does not tell us anything about their mode of action. Thus, it is common to refer to these additives as either UV absorbers or free-radical scavengers. Later in this chapter, these four light stabilizer classes will be discussed in greater detail.

3.1 UV absorbent pigments

Incorporating pigments is probably the oldest form of providing UV protection. Titanium dioxide and carbon black are both capable of absorbing UV light and thus help to stabilize paint films [24, 25]. Obviously pigments are not universal UV absorbers due to their colour. Furthermore, pigments such as titanium dioxide can also cause photo-oxidative degradation of polymers. The latter is available in various forms, namely anatase (treated or untreated) and rutile (treated or untreated) and depending on its form and surface treatment, it can initiate polymer degradation through the formation of hydroxyl and hydroperoxide radicals [26–28].

Figure 3.2 describes the cyclic process of radical formation. Generally, hydroxyl groups are present on the TiO_2 surface. Absorption of UV light causes the formation of a positively charged hole on the surface, which immediately reacts with the hydroxyl group to furnish a hydroxyl radical (Equations 3.1 and 3.2). The release of an electron through reaction (3.1) results in the reduction of Ti^{4+} to Ti^{3+} (Equation 3.3).

Oxidation of Ti^{3+} by atmospheric oxygen generates Ti^{4+} along with the adsorbed oxygen (Equation 3.4). These species along with hydrogen ions formed during the dissociation of water provide hydroperoxide radicals according to Equation 3.5. The cyclic process then starts new via Equation 3.6.

Since pigments can only serve as UV absorbers under certain conditions, the search began for colourless molecules capable of absorbing UV light without undesirable side effects.

3.2 UV absorbers

As already pointed out, the main function of UV absorbers is to absorb UV radiation in competition with the chromophore (Ch) present in the polymer, thereby filtering out the UV light that is harmful to the polymer, before Ch* has had a chance of forming (see Figure 3.1). Table 3.1 gives an overview of the spectral regions at which polymers exhibit maximum sensitivity [22, 29, 30].

Note that there is some slight variation in the absolute ranges in the technical literature.

Table 3.1: Wavelengths of maximum sensitivity of polymer systems [22, 29, 30]

Polymer system	Wavelengths of max. sensitivity [nm]
Polyethylene	300
Polypropylene	310
Polystyrene	320
Polyvinyl chloride	310
Polycarbonate	295, 345
Polymethyl methacrylate	290 to 315
Polyester	315 to 325
Celulose acetate butyrate	295, 298
Oils and alkyd resins	280 to 310

Based on the figures in Table 3.1, the UV absorber must absorb light in the region between 290 and 350 nm. If, however, possible impurities unavoidable in industrially produced polymers, as well as additives [31, 32], pigments, fillers or even dyes are taken into account, its absorption should extend above 350 nm [33], without adversely affecting the colour of the coating.

$$(Ti^{4+} \cdot OH^-) \xrightarrow{h\nu} (Ti^{4+} \cdot OH)^+ + e^- \quad \text{(Eq. 3.1)}$$

$$(Ti^{4+} \cdot OH)^+ \longrightarrow Ti^{4+} + HO\cdot \quad \text{(Eq. 3.2)}$$

$$Ti^{4+} + e^- \longrightarrow Ti^{3+} \quad \text{(Eq. 3.3)}$$

$$Ti^{3+} + O_2 \longrightarrow (Ti^{4+} \cdot O_2) \quad \text{(Eq. 3.4)}$$

$$(Ti^{4+} \cdot O_2) + H^+ \longrightarrow Ti^{4+} + HOO\cdot \quad \text{(Eq. 3.5)}$$

$$Ti^{4+} + OH^- \longrightarrow (Ti^{4+} \cdot OH^-) \quad \text{(Eq. 3.6)}$$

Figure 3.2: Photo-chemical reactions on the surface of TiO_2 particles [26, 27]

The role of UV absorbers is to absorb harmful UV light and quickly transform it into harmless heat. During this process, the excited state energy of the molecule is converted into vibrational and rotational energy. It is essential that the dissipation of the energy of the UV absorber takes place more rapidly than the any reaction within the substrate and that neither the UV absorber nor the polymer is damaged during energy conversion.

3.2.1 UV absorber classes

Various organic molecules are capable of absorbing harmful UV light and converting it into harmless heat. These include [21, 31, 34–39]:

a. 2-(2-hydroxyphenyl)-benzotriazoles
b. 2-hydroxybenzophenones
c. hydroxyphenyl-s-triazines
d. oxanilides
e. hydroxyphenylpyrimidines
f. salicylic acid derivatives
g. cyanoacrylates

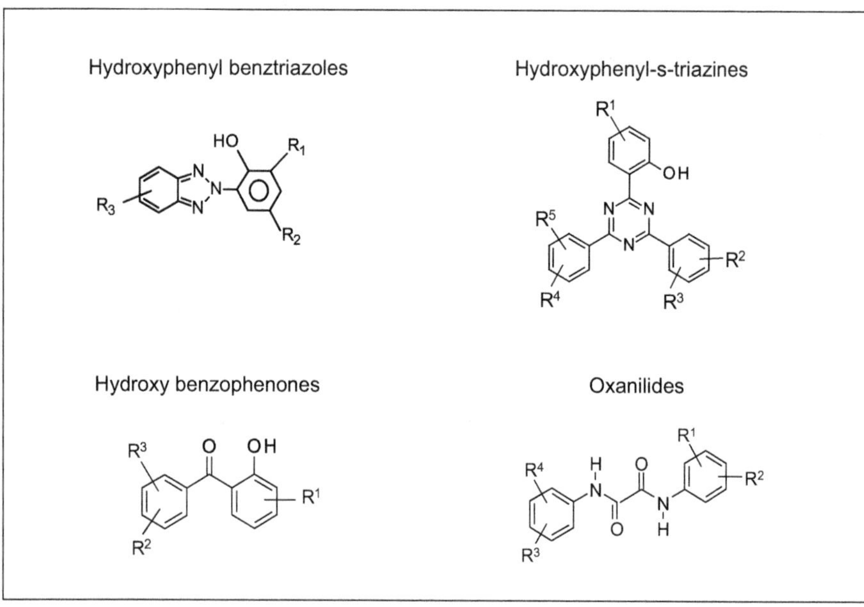

Figure 3.3: The four most important UV absorber classes

The most important of classes of UV absorbers are a) to d), whereas 2-(2-hydroxyphenyl)-benzotriazoles are most widely used for polymers and paints. The chemical structures of these important classes of UV absorbers are shown in Figure 3.3.

For simplicity, 2-(2-hydroxyphenyl)-benzotriazoles will be referred to as *benzotriazoles (BTZ)*, 2-hydroxybenzophenones as *benzophenones (BP)* and hydroxyphenyl-s-triazines as *HPT.*

Each of these types of UV absorber classes is characterized by a typical absorption spectrum (Figure 3.4) and transmittance spectrum (Figure 3.5).

UV absorbers with specific substitution patterns leading to a different shape or a shift of the absorption spectrum towards the visible (red shifted UV absorbers) will be discussed separately in Section 3.2.3.

Figure 3.4 shows that:

- oxanilides have an absorption maximum at about 300 nm in the range of 280 to 400 nm
- conventional benzotriazoles, hydroxyphenyl-s-triazines and benzophenones have two absorption maxima (one at shorter wavelengths at about 300 nm and another at longer wavelengths above 320 nm)

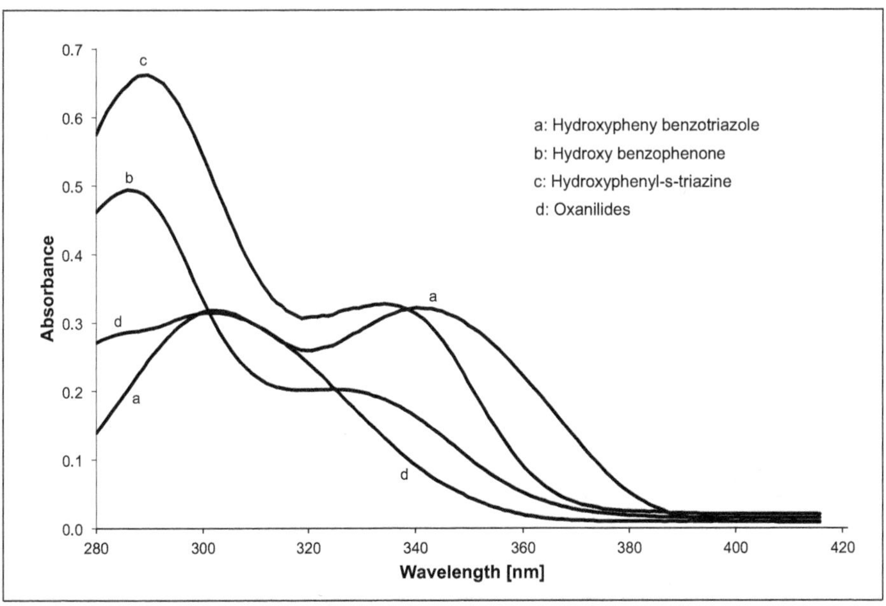

Figure 3.4: Absorption spectra of different conventional UV absorber classes c = 10 mg/L in chloroform (1 cm cell)

- hydroxyphenyl-s-triazines have the strongest absorption in the region of 300 nm, followed by benzotriazoles and benzophenones.
- The long wavelength absorption of benzotriazoles occurs above 340 nm, that of hydroxyphenyl-s-triazines at 335 to 340 nm. The corresponding absorption of benzophenones is observed at 320 to 330 nm.

The exact positions of the absorption maxima, especially the longer-wavelength absorptions, ones – depend on the substituents of the molecules (see Figure 3.3) and the medium in which the spectrum is measured.

The transmittance spectra of conventional UV absorbers are shown in Figure 3.5.

Generally, the more harmful UV light can be filtered out, the farther will the absorption edge reach into the long-wave UV region. However, the perfect UV absorber must have a little transmittance as possible in the UV but at the same time perfect transmittance in the visible light, above 400 nm, to avoid discolouration of the coating.

Although it has proven difficult to extend the absorption of conventional benzotriazoles or hydroxyphenl-s-triazines further into the long-wave UV without causing discolouration, surprising effects can be achieved by using blends of different types of UV absorbers [40].

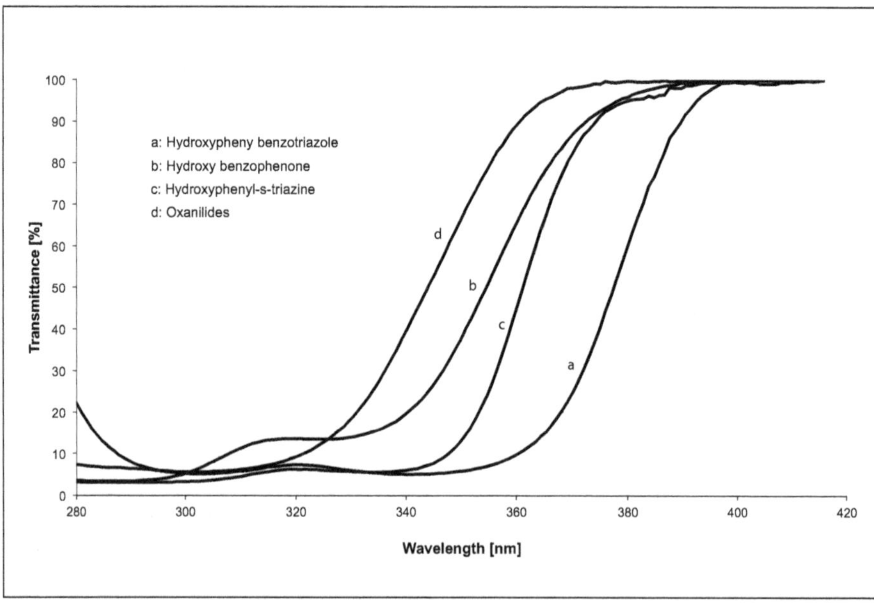

Figure 3.5: Transmittance spectra of different conventional UV absorber classes c = 10 mg/L in chloroform (1 cm cell)

The effectiveness of an UV absorber can be quantified by the Lambert-Beer Law:

$E = Abs = \log I_o / I = \varepsilon \cdot c \cdot d$

where E = extinction
Abs = absorbance (which is not the same as absorption!)
I_o = intensity of incident light
I = intensity of emergent light
ε = extinction coefficient [l/mol cm]
c = concentration [mol/l]
d = thickness (film, substrate, cell) [cm]

The extinction depends on the wavelength and may be regarded as a measure of the stabilizing or screening effect of the UV absorber. In other words, the higher the extinction, the more UV light absorbed and the greater the stabilizing effect – provided the UV absorber itself is not destroyed by the absorption of the light. The magnitude of extinction depends on the extinction coefficient ε, the concentration c of the UV absorber in the polymer and the film thickness d of the unpigmented polymer.

The extinction coefficient ε is specific for each molecule and wavelength dependent, i.e. it is a constant for every given UV absorber. In the case of benzotriazoles, hydroxyphenyl-s-triazines and benzophenones, the extinction coefficient of the long-wave absorption maximum is determinant. This means that the stabilizing effect of a given UV absorber can be altered by varying the concentration c and/or the film thickness d as shown in the following example.

Let us assume that a UV absorber UVA-X has an extinction coefficient of 15.000 L/mol cm at 345 nm, its concentration is 1.5×10^{-4} mol/L and the polymer film thickness is 0.1 cm. This will produce an extinction of 0.225. An identical stabilizing effect arises, for example, by cutting the concentration in half and doubling the film thickness. To increase the stabilizing effect, an increase either the concentration c or the film thickness d is necessary.

3.2.2 Mode of action of UV absorbers

The mode of action of UV absorbers depends on structure. Those that contain phenolic groups such as benzotriazoles, benzophenones and hydroxyphenyl-s-triazines dissipate energy differently than those without, such as oxanilides. An effective UV absorber must

- absorb UV light faster and more readily than the polymer it is meant to stabilize
- dissipate the absorbed energy before unwelcome side reactions can occur.

This means that transformation of the energy absorbed in the form of UV light must take place in the singlet state. Inter-system crossing (transition $S_1 \rightarrow T_1$) and therefore phosphorescence must be excluded [21].

3.2.2.1 Phenolic UV absorbers

The mechanism of energy conversion in the case of benzotriazoles and benzophenones was the subject of exhaustive investigations already in the late sixties and early seventies [41–46]. Figure 3.6 shows the energy conversion for benzotriazole.

Once the benzotriazole molecule has absorbed UV light, proton transfer from oxygen to the nitrogen occurs in the excited singlet state. This type of proton transfer was first observed in salicylic acid esters [47]. The photo-tautomer reverts to the original molecule via rapid, radiation-less deactivation, whereupon the cycle can repeat. This process is usually referred to as **keto-enol tautomerism.**

The phenolic hydroxyl group and an intra-molecular hydrogen bridge are absolutely essential for keto-enol tautomerism. This interaction is responsible for the geometry of the entire molecule [45, 46]. An intact intra-molecular hydrogen bridge favours a planar structure, whilst molecules lacking intra-molecular hydrogen bridges, e.g. when hydrogen is substituted by methyl, assume a twisted geometry characterized by a dihedral angle of

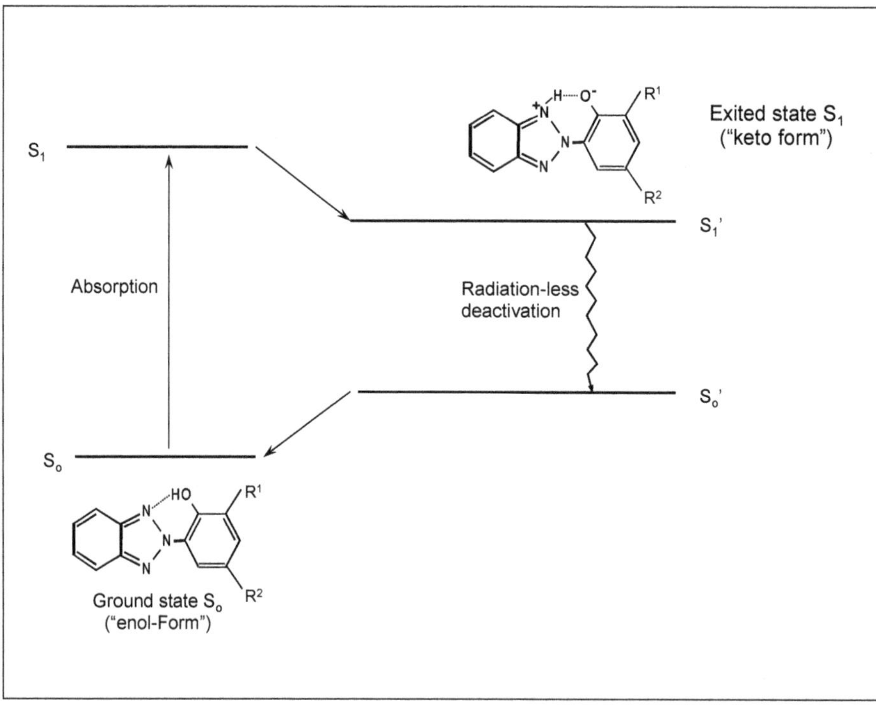

Figure 3.6: Energy conversion of phenolic UV absorbers, using benzotriazoles as example according to Otterstedt [44]

~56° [45]. Intra-molecular hydrogen bridges are also sensitive to the medium (solvent, polymer, etc.) [46, 48]. In polar medium, an *inter*-molecular hydrogen bridge with the medium may be preferred to the *intra*-molecular hydrogen bridge.

With an intact intra-molecular hydrogen bridge, hydrogen transfer takes place rapidly ($k_{s1s1'} > 10^{11}\ s^{-1}$). As this process is prevented when an inter-molecular hydrogen bridge is present, undesirable reactions can occur, for example through the triplet state intermediate that has a comparatively long lifetime.

As aforementioned in Section 2.1.1, many photo-chemical reactions arise from the excited triplet state. This is also true for the phenolic UV absorbers where the intermediacy of the triplet state can lead to its decomposition. Furthermore, the interaction of a UV absorber in the triplet state with oxygen can give rise to singlet oxygen, a very strong oxidant that can degrade polymers [47].

Blocking of the hydroxyl group by a methyl group is irreversible and results in the total loss of activity [46]. However, blocking the hydroxyl group of a benzotriazole with an

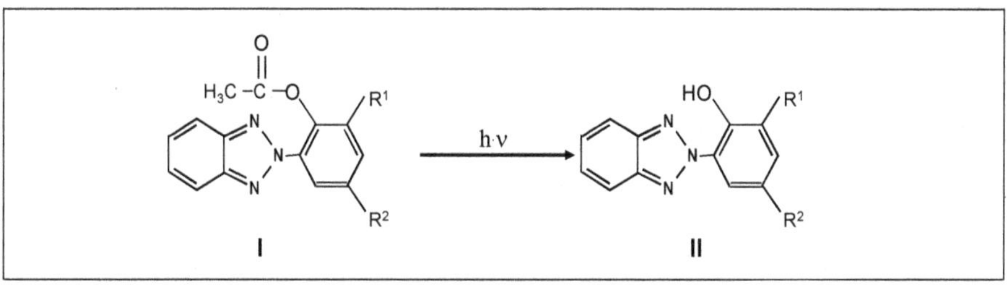

Figure 3.7: Deacetylation of a blocked benzotriazole (I) through irradiation [49, 50]

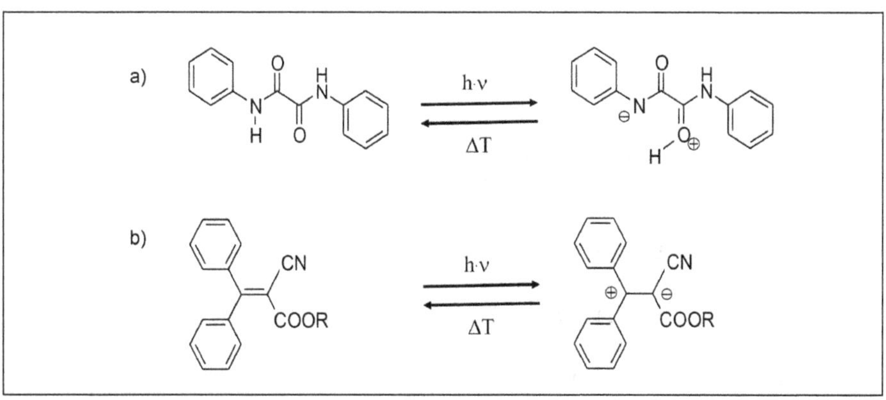

Figure 3.8: Energy conversion of non-phenolic UV absorbers;
a) oxanilides [38, 55], b) cyanoacrylates [38]

acetate group (Figure 3.7) is reversible under the influence of UV light and leads to the regeneration of the active UV absorber [49, 50].

The intact intra-molecular hydrogen bridge in hydroxyphenyl-benzotriazoles gives rise to the long-wave absorption with a maximum from 340 to 350 nm (Figure 3.4). The exact position of the long-wave absorption maximum depends upon the polarity of the medium in which the spectrum has been recorded and on the substituents, especially those on the benzotriazole moiety [48, 51]. The intra-molecular hydrogen bridge has no effect on the short-wave absorption at approximately 300 nm. The same is true for the hydroxyphenyl-s-triazines [52–54].

3.2.2.2 Non-phenolic UV absorbers

Of the non-phenolic UV absorbers, the oxanilides are most often employed in coatings. The absorption maximum of this class of UV absorbers occurs around 300 nm (Figure 3.4). Since the oxanilides do not contain phenolic hydroxyl groups, the wavelength of maximum absorption is independent of the medium. The exact fate of the absorbed energy has not been clearly elucidated but experimental results indicate an intra-molecular proton transfer [38, 55] (Figure 3.8).

3.2.3 Examples of UV absorbers

The specific UV absorbers that will be discussed in the subsequent chapters and sections of this book are listed below with their common names together with their proper chemical names. Figure 3.9 to 3.12 show the chemical structures of these substances.

a. **Hydroxyphenyl-benzotriazole UV absorbers (Figure 3.9)**
   - benzotriazole-1 (BTZ-1): 2-(benzotriazole-2-yl)-4-methylphenol
   - benzotriazole-2 (BTZ-2): 2-(benzotriazole-2-yl)-4,6-bis-(2-methylbutan-2-yl)phenol
   - benzotriazole-3 (BTZ-3): 2-(benzotriazole-2-yl)-4,6-bis-(2-phenylpropan-2-yl) phenol
   - benzotriazole-4 (BTZ-4): 2-(benzotriazole-2-yl)-6-(2-phenylpropan-2-yl)-4-(2,4,4-trimethyl-pentan-2-yl)phenol
   - benzotriazole-5 (BTZ-5): 2-tert-butyl-6-(5-chloro-2H-benzotriazole-2-yl)-4-methylphenol
   - benzotriazole-6 (BTZ-6): 6-butyl-2-[2-hydroxy-3-(1-methyl-1-phenylethyl)-5-(1,1,3,3-tetramethylbutyl)phenyl]-pyrrolo[3,4-f]benzotriazole-5,7(2H,6H)
   - benzotriazole-7 (BTZ-7): reaction product of 2-(2-hydroxy-3-tert-butyl-5-methylpropionat)-2H-benzotriazole and PEG 300
   - benzotriazole-8 (BTZ-8): 4-methylhexyl-3[3-(benzotriazole-2-yl)-t-tert-butyl-4-hydroxyl-phenyl)propanoat

b. Hydroxyphenyl-s-triazine UV absorbers (Figure 3.10a and Figure 3.10b)

- HPT-1: reaction products of 2-(4,6-bis(2,4-dimethylphenyl)-1,3,5-triazine-2-yl)-5-hydroxyphenol with ((C12-C13 alkoxy)methyl)oxyran
- HPT-2: 6-[2,6-bis(2,4-dimethylphenyl)-1H-1,3,5-triazine-4-ylidene]-3-[3-(ethylhexoxy)-2-hydroxy-propoxy]cyclohexa-2,4-dien-1-one
- HPT-3: 2-(octyloxy-2-hydroxyphenyl)-4,6-bis(2,4-dimethylphenyl)-1,3,5-triazine
- HPT-4: 6-methylheptyl-2-[4-[4,6-bis(4-phenylphenyl-1H-1,3,5-triazine-2-ylidene]-3-oxocyclohexa-1,5-dien-1-yl]oxypropanoat
- HPT-5: 3-butoxy-6-[4-butoxy-6-oxocyclohexa-2,4-dien-1-ylidene)-6-(2,4-dibutoxyphenyl)-1H-1,3,5-triazine-2-ylidene]cyclohexa-2,4-dien-1-one
- HPT-6: reaction products of bromo-propionic acid isooctylesters and 2,4,6-tris[dihydroxy-phenyl]-1,3,5-triazine

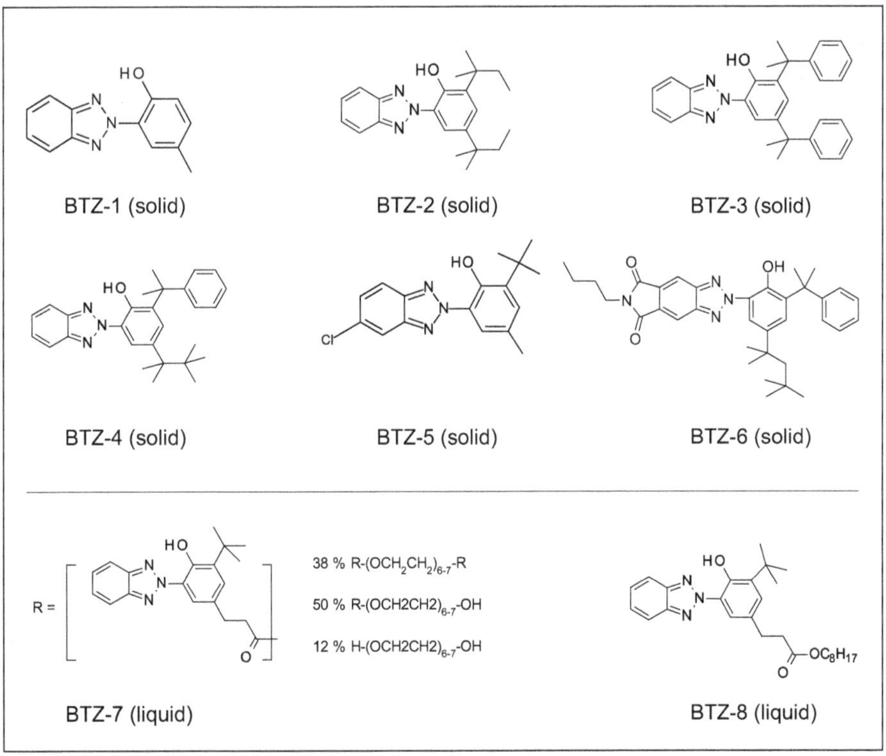

Figure 3.9: Examples of BTZ (benzotriazole) UV absorbers

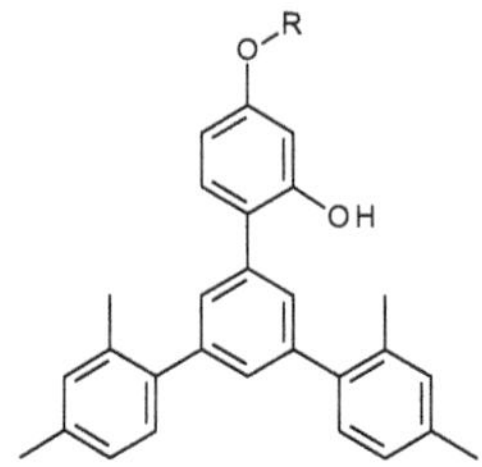

Hydroxypenyl-s-triazine (HPT)

HPT-1: R = CH_2-CH(OH)-CH_2-O-CH-$C_{12}H_{25}$/C_{13}/H_{27} (liquid)

HPT-2: R = CH_2-CH(OH)-CH_2-O-CH(C_2H_5)(C_4H_9) (solid)

HPT-3: R = C_8H_{17} (solid)

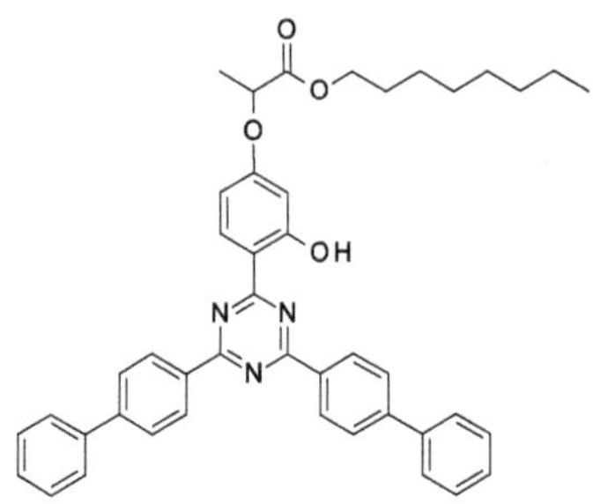

HPT-4 (solid)

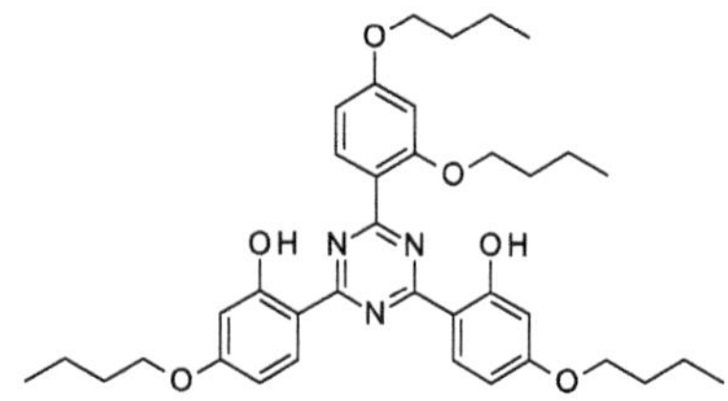

HPT-5 (solid)

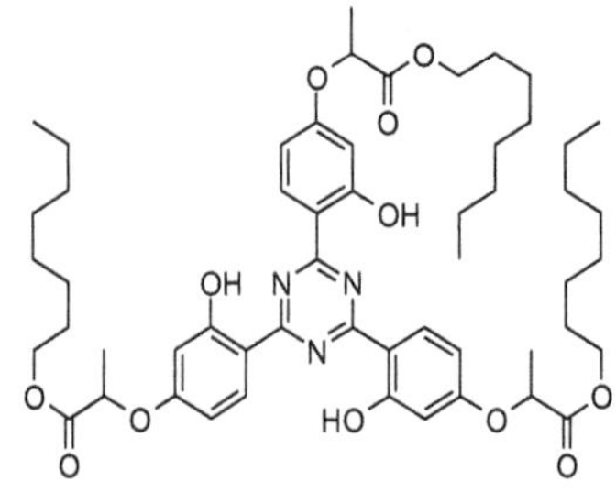

HPT-6 (liquid)

Figure 3.10a (above) and b: Examples of HPT (hydroxyphenyl-s-triazine) UV absorbers

c. Hydroxybenzophenone UV absorbers (Figure 3.11)
- benzophenone-1 (BP-1): 2,4-dihydroxy-benzophenone
- benzophenone-2 (BP-2): 2-hydroxy-4-octyloxybenzophenone
- benzophenone-3 (BP-3): 2-hydroxy-4-dodecyloxybenzophenone

d. Oxanilide UV absorbers (Figure 3.12)
- oxanilide-1: N-(2-ethoxyphenyl)-N'-(2-ethylphenyl)-ethanediamide
- oxanilide-2: N-(2-ethoxyphenyl)-N'-(4-isododecylphenyl)-ethandiamide

It was stated earlier that extension of the long-wave absorption of benzotriazoles or hydroxyphenyl-s-triazine may give rise to colour. There are, however, applications where a (slight) colouration of the UV absorber might be acceptable. Figure 3.13 illustrates the effect of certain substituents on the position of the absorption maximum for benzotriazole based UV absorbers. Halogenation of the benzotriazole ring (BTZ-5) causes a shift of the long-wave maximum of about 10 nm. If the benzene ring of the benzotriazole is integrated into a phthalic imide system (BTZ-6) [56], for example, a more significant shift of around 30 nm is achieved. Figure 3.14 depicts similar absorption shifts for the hydroxyphenyl-triazines. In the series of UV absorbers HPT-1, HPT-4, HPT-5 and HPT-6, a sequential shift of the absorption maximum is clearly seen as well as extended tailing into the visible region above 400 nm.

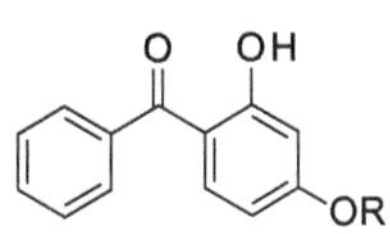

Hydroxy benzophenone

BP-1: R = H (solid)

BP-2: R = C_8H_{17} (solid)

BP-3: R = $C_{12}H_{25}$ (solid)

Figure 3.11: Examples of BP (benzophenone) UV absorbers

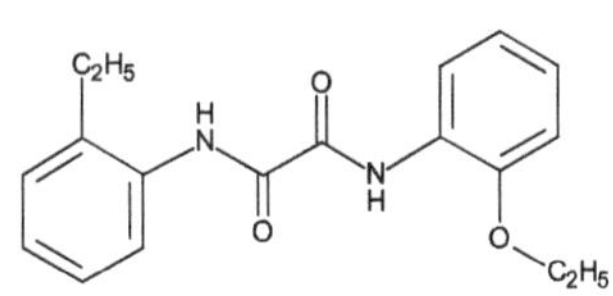

Oxanilide-1 (solid)

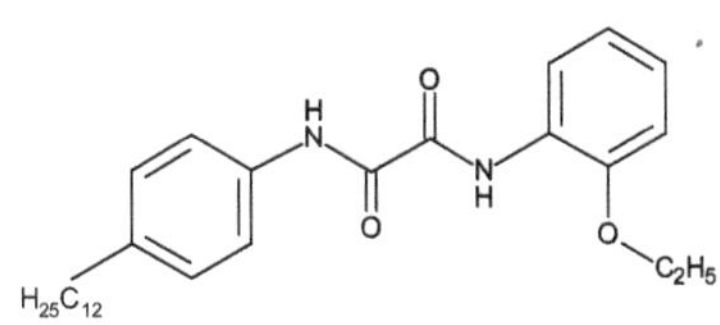

Oxanilide-2 (liquid)

Figure 3.12: Examples of oxanilide UV absorbers

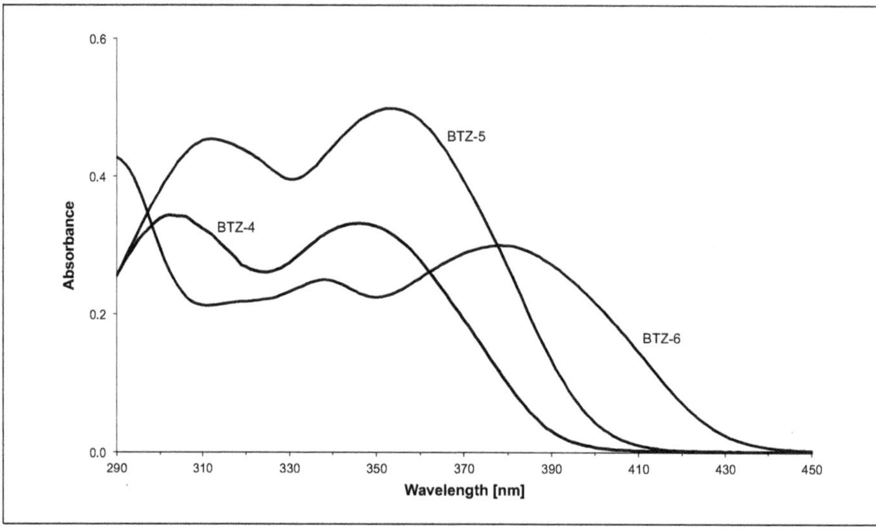

Figure 3.13: Influence of substituent pattern on absorption in case of BTZ UV absorbers (c = 10 mg/L in chloroform)

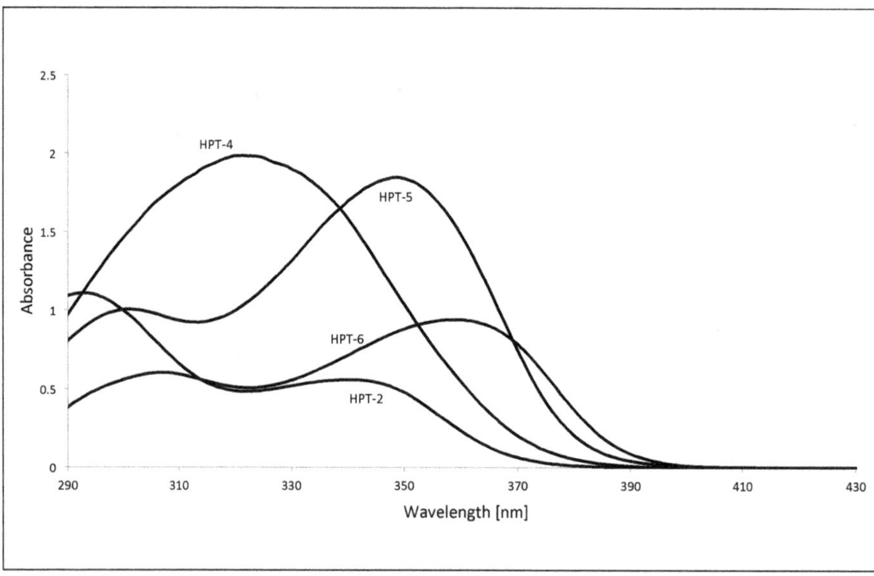

Figure 3.14: Influence of substituent pattern on absorption in case of HPT UV absorbers (c = 20 mg/L in toluene)

Table 3.2: Wavelengths of maximum long-wave absorption (λ_{max}) of different UV absorbers and the associated extinction coefficients ε. Solvent: chloroform, concentration = 1.4 x 10^{-4} mol/L; 1 cm cell

UV absorber	λ max [nm]	ε [L · Mol^{-1} · cm^{-1}]
BTZ-1	341	~ 16,100
BTZ-2	346	~ 15,600
BTZ-6	378	~ 16,900
BTZ-8	344	~ 15,400
HPT-1	338	~ 19,400
HPT-3	339	~ 21,700
HPT-4	322	~ 67,500
HPT-5	348	~ 58,400
HPT-6	359	~ 13,300
BP-2	327	~ 10,700
BP-3	327	~ 10,600
Oxanilide-1	301	~ 14,500

Such "red-shifted" UV absorbers provide more protection in the border region between UV and VIS light but cause a (slight) yellowing of the coating.

Table 3.2 gives an overview of the wavelengths for maximum **long-wave** absorption (λ_{max}) for the different UV absorbers, as well as the extinction coefficients ε at those wavelengths. As seen in Table 3.2, the position of the long-wave absorption maximum of a UV absorber depends on structure and the extinction coefficient is a function of the molecular weight and the substitution pattern.

UV absorbers for the polymers commonly used today must have a long-wave absorption maximum between 330 and 350 nm. Molecules with long-wave absorption maximum at shorter wavelengths (shift towards blue) would not provide sufficient stabilization since there would be too little absorption at wavelengths approaching 380 nm (see also the transmission spectra in Figure 3.4). A λ_{max} of significantly more than 350 nm, however, may possess a yellow tinge and adversely affect the colour of the coating as mentioned above. With the exception of specialized applications requiring broader spectral coverage near or above 400 nm, the bulk of the market relevant coatings can be aptly protected by conventional hydroxyphenyl benzotriazole and hydroxyphenyl-s-triazine UV absorbers.

3.3 Free-radical scavengers

At the beginning of Section 3 we stated that the excited chromophore Ch^* can either decompose to form radicals which can then react with the polymer and/or atmospheric oxygen, or remove a hydrogen atom directly from the polymer to produce a radical reaction (see Figure 3.1). To suppress the ensuing radical reactions, molecules are required that can trap the radicals formed, namely free-radical scavengers, thereby interrupting these chain reactions. The two most important classes of radical scavengers are **antioxidants** and **sterically hindered amines.**

The importance of free-radical scavengers was recognized and discussed as early as the mid-1960s [57–59].

3.3.1 Antioxidants

Antioxidants can be divided into two groups according to their mode of action:
1. Compounds that interrupt the chain reaction (kinetic mode of action)
2. Compounds that function as peroxide decomposing agents (ionic mechanism).

The first group of antioxidants are also termed **primary antioxidants** [60] whereas those of the second group are referred to as secondary antioxidants. This section deals with the first group of antioxidants, typified by the sterically hindered phenols (see Figure 3.15). The secondary antioxidants are discussed in Section 3.5.

Figure 3.16 shows the mode of action of primary antioxidants, as understood today. Many of the individual reactions are not fully understood and depend greatly upon the polymer, additives, fillers, pigments, impurities and temperature.

The stability of the phenoxy radical formed according to equation (a) depends on the substituents R_1-R_3 and particularly on the amount of resonance stabilization (delocalization of the electron).

The more stable the phenoxy radical (Figure 3.17), the less likely it is to initiate further chain reactions. The size of the substituents R_1-R_3 has a great effect on the fate of the phenoxy radical. If, for example, R_1 is hydrogen and R_3 is a methyl group or both R_1 and R_2 or R_1 and R_3 are alkyl groups, dimerization in the **para**-position relative to the oxygen atom is probable. This leads to the formation of a cyclohexanone dimer [12]. As a general rule, R_1 and R_2 should be fairly bulky groups such as tertiary butyl, whereas R_3 can be used to control the compatibility with the polymer.

The presence and concentration of phenoxy radicals can be determined by ESR (electron spin resonance) spectroscopy [62]. Phenolic antioxidants are used especially as stabilizers against thermo-oxidation, e.g. at high processing temperatures.

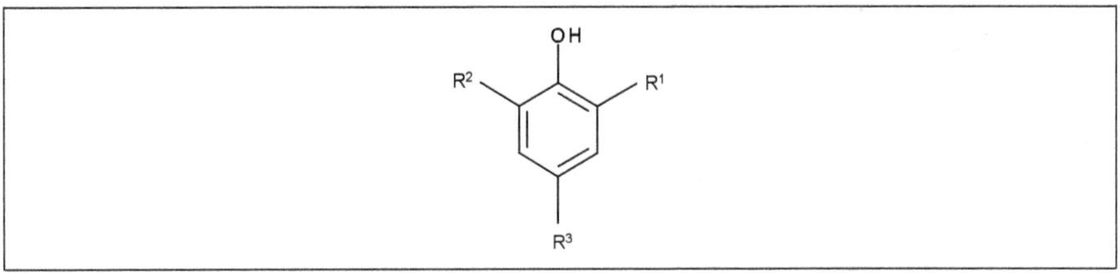

Figure 3.16: Mode of action of sterically hindered phenols [60]

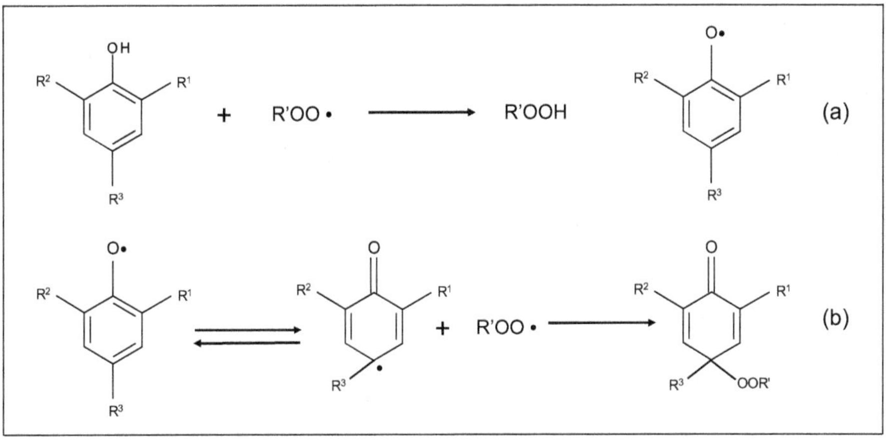

Figure 3.15: General structure of sterically hindered phenols (phenolic antioxidants)

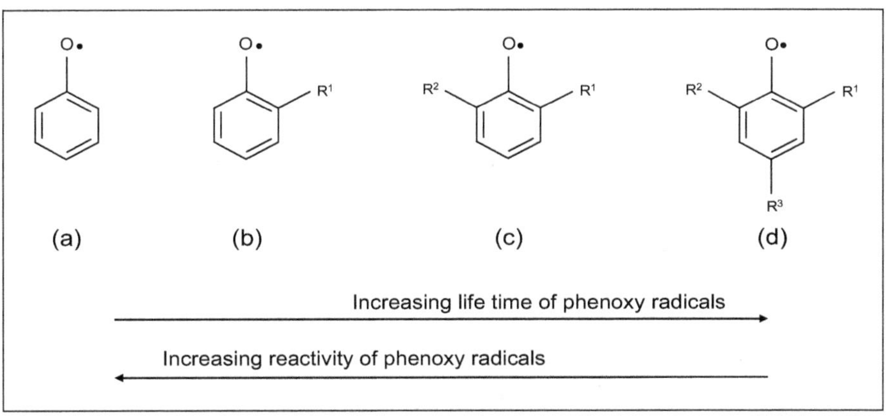

Figure 3.17: Dependence of life time and reactivity of phenoxy radicals on substitution [61]

One important drawback of phenolic antioxidants is their **non-cyclic** mode of action. This means, in effect, that after a certain period – depending on the initial concentration and conditions within the polymer – no more antioxidant is left to prevent undesirable free-radical reactions. Therefore, a preferred antioxidant is one that exhibits a **cyclic** mode of action and remains effective for an almost unlimited period of time.

3.3.2 Sterically hindered amines

Sterically hindered amines have been used to stabilize commercial polymers since the early 1970s. These materials are generally referred to as HALS, an acronym for hindered amine light stabilizers and the most common are derivatives of 2,2,6,6-tetramethylpiperidine (Figure 3.18). This abbreviation HALS is also used throughout this book.

ESR spectroscopy has shown that, under photo-oxidative conditions, HALS are converted into the corresponding stable nitroxyl radical, at least to a large extent [63, 64]. Stable nitroxyl radicals have been known since 1845, when *Fremy* for the first time synthesized a stable inorganic nitroxyl radical [65]. In 1901, a stable organic nitroxyl radical was synthesized for the first time [66]. In the years that followed, a variety of stable diaryl and alkylaryl nitroxyl radicals were prepared [67–69].

Nitroxyl radicals are stable only if they have no active substituents, such as hydrogen atoms, in the alpha position. When a hydrogen atom is present in the alpha position, disproportionation of the molecule into a nitron and a hydroxylamine occurs as shown schematically in Figure 3.19 [70].

The synthesis of 2,2,6,6-tetramethyl-4-oxo-piperidine-N-oxyl radical by *Lebedev* in 1961 [71] was the starting point for the piperidine-based nitroxyl. *Rozantsev* published a paper in 1970 [72] on 4-amino-2,2,6,6-tetramethylpiperidine-N-oxyl radical and its derivatives.

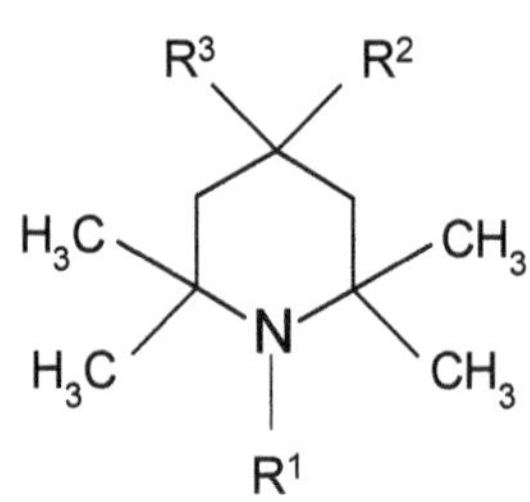

Figure 3.18: General structure of sterically hindered amines based on 2,2,6,6-tetramethyl piperidine

Figure 3.20 depicts two compounds on which most of today's commercially important products are based.

Table 3.3 (see next page) shows some of the most important HALS used on an industrial scale today that will be discussed in detail in the following chapters.

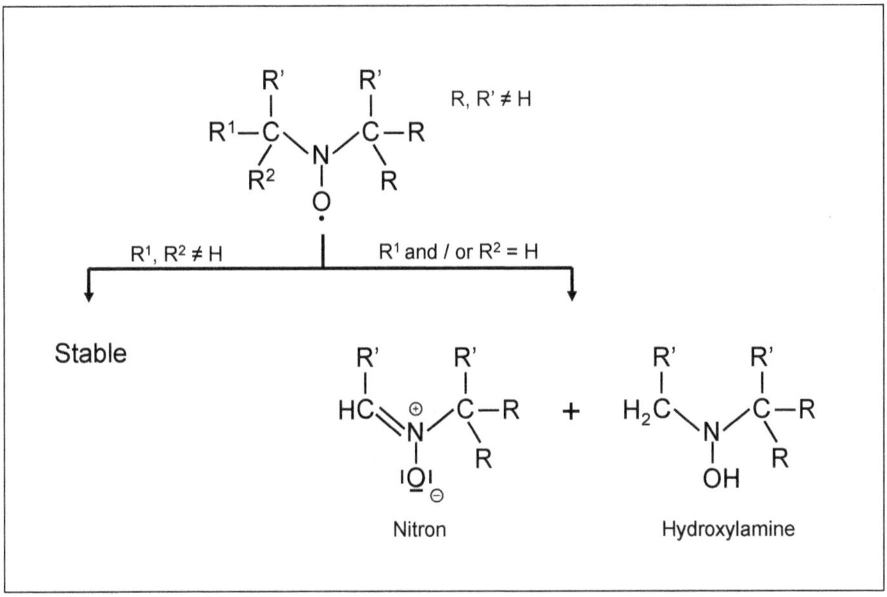

Figure 3.19: Effect of substitution in the alpha position to the nitroxyl group on the stability of nitroxyl radicals [70]

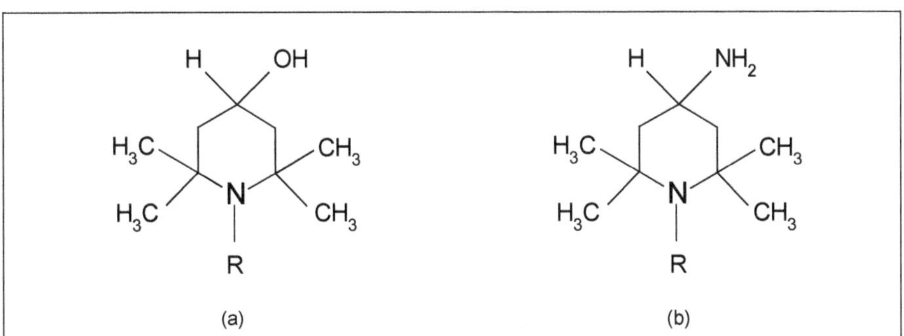

Figure 3.20: 4-hydroxy- (a) and 4-amino- (b) tetramethyl piperidine

Table 3.3: Structures of technically important HALS

Name	Structure	
HALS-1 HALS-2 HALS-3	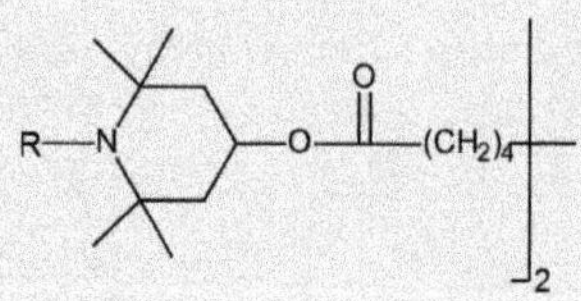	R = H (solid) R = CH_3 (liquid) R = C_8H_{17} (liquid)
HALS-4	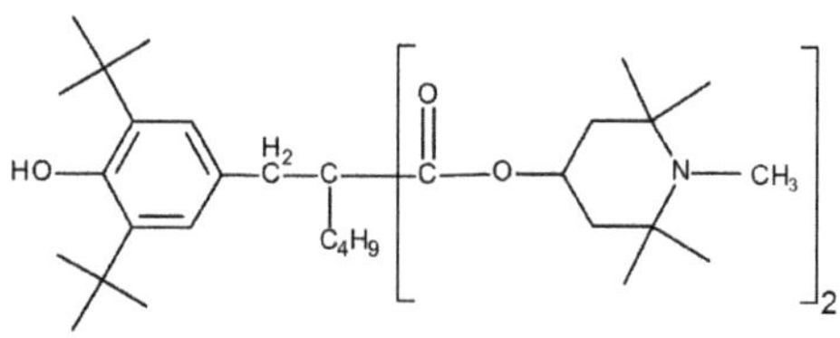	(solid)
HALS-5	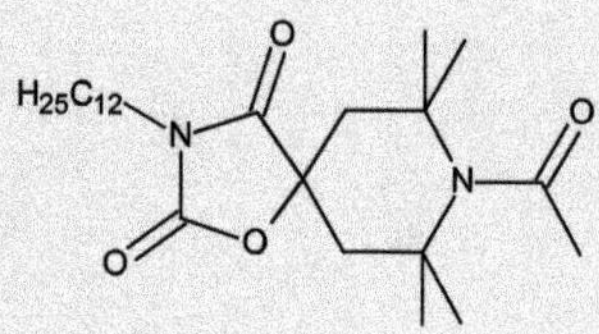	(solid)
HALS-6	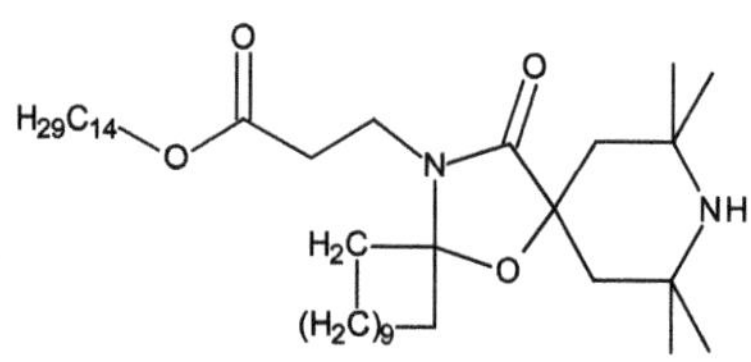	(liquid)
HALS-7	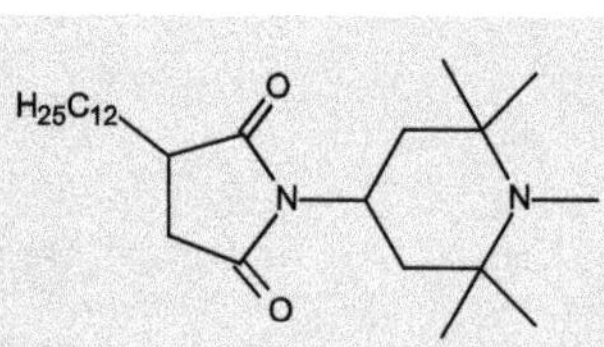	(liquid)
HALS-8	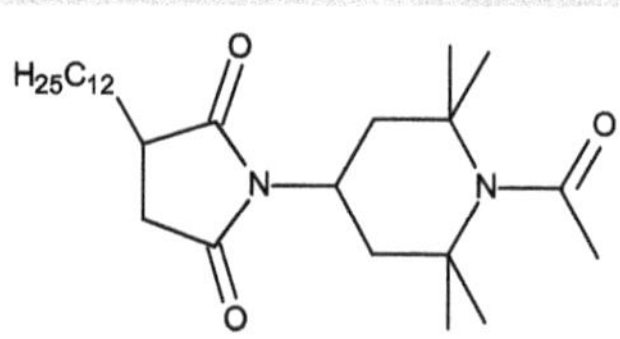	(liquid)

Table 3.3: Structures of technically important HALS (continuation)

Name	Structure
HALS-9	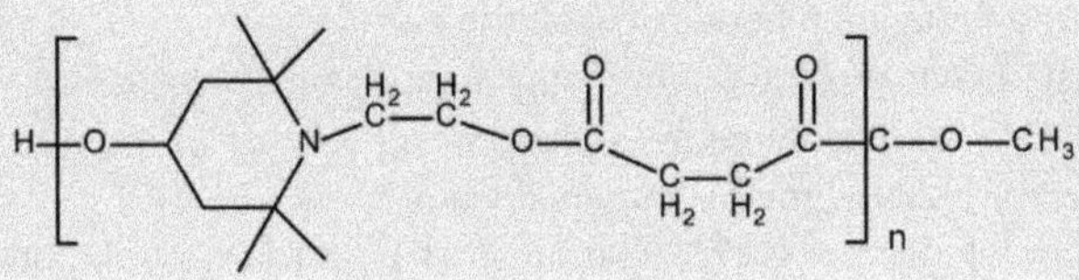 (solid)
HALS-10	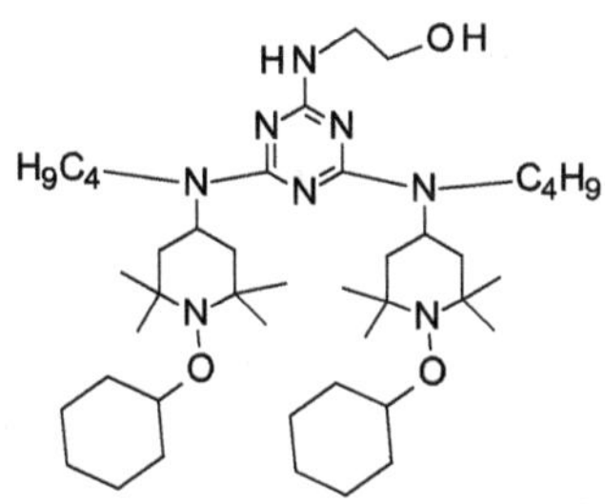 (liquid)
HALS-11	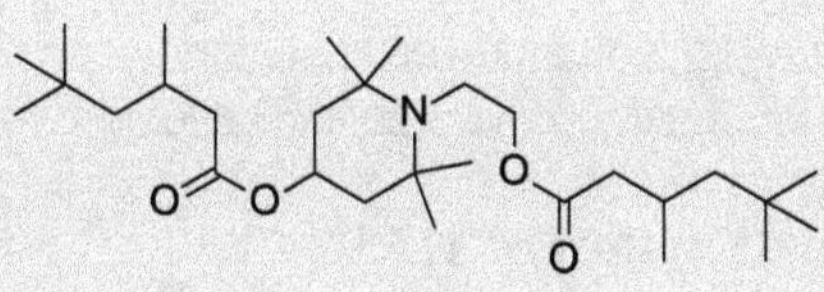 (liquid)
HALS-12	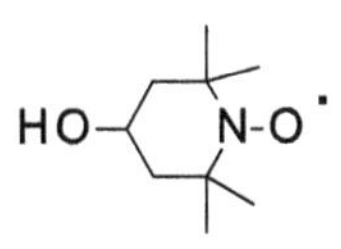 (solid)
HALS-13	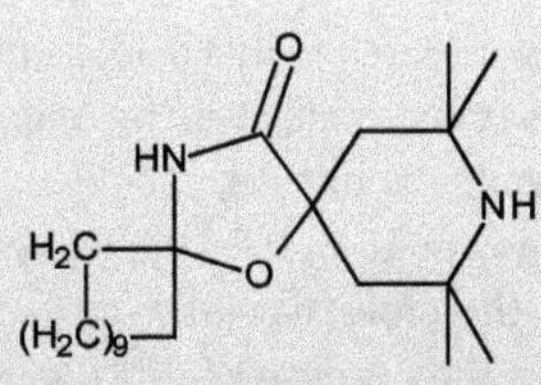

The chemical names for these HALS are as follows:

HALS-1:	bis(2,2,6,6-tertamethyl-4-piperidinyl)sebacate
HALS-2:	bis(2,2,6,6-pentamethyl-4-piperidinyl)sebacate + methyl(1,2,2,6,6-pentamethyl- 4-piperidinyl)sebacate
HALS-3:	bis(1-octyloxy-2,2,6,6-tetramethyl-4-piperidinyl)sebacate
HALS-4:	bis(1,2,2,6,6-pentamethyl-4-piperidinyl)[[3,5-bis(1,1-dimethylethyl)-4-hydroxy-phenyl]methyl]butylmalonate
HALS-5:	8-acetyl-3-dodecyl-7,7,9,9-tetramethyl-1,3,8-triazaspiro[4,5]decane-2,4-dione
HALS-6:	7-Oxa-3,20-diazadispiro5.1.11.2heneicosane-20-propanoic acid, 2,2,4,4-tetramethyl-21-oxo-, dodecyl/tetradecyl ester
HALS-7:	2-dodecyl-N-(2,2,6,6-tetrmethyl-4-piperidinyl)succinimide
HALS-8:	N-(1-acetyl-2,2,6,6-tetramethyl-4-piperidinyl)-2-dodecylsuccinimide
HALS-9:	poly(4-hydroxy-2,2,6,6-tetramethyl-1-piperidine ethanol-alt-1,4-butanedioic acid)
HALS-10:	ethanol, 2-amino, reaction products with cyclohexane and peroxidized N-butyl-2,2,6,6-tetramethyl-4-piperidinamine-2,4,6-trichloro-1,3,5-triazine reaction products
HALS-11:	2,2,6,6-tetrametyl-1-[2-(3,5,5-trimethyl-hexanoyloxy-)ethyl]-piperidine-4-yl ester
HALS-12:	4-Hydroxy-2,2,6,6-tetramethylpiperidinoxyl
HALS-13:	2,2,4,4-tetramethyl-7-oxa-3,20-diazadispiro[5.1.11.2]-henicosan-21-one

3.3.2.1 Mode of action of HALS

Probably the most familiar depiction of the mode of action of HALS is the so-called "**Denisov cycle**" shown schematically in Figure 3.21 [73]. Equation (a) represents the transformation of HALS into the corresponding nitroxyl radical [63, 64] whereas equations (b) and (c) make up the actual cycle, the reactions and regeneration of the nitroxyl radical.

Even today, years after the initial proposal by Denisov, the formation of >NO· (Equation (a) in Figure 3.21) and the inter-mediacy of >NOR (Equation (b)) remain uncontested [74].

The "*Denisov cycle*" is based on experiments performed in polyolefins, a relatively easy substrate to investigate in comparison to typical coatings. In spite of the great number of publications and continuing research in this field, a universal mode of action of HALS valid for all polymeric substrates remains elusive. The major challenge is that most of the intermediates formed are not stable and, thus, cannot be isolated and analysed. Most of the scientific literature concerns liquid or solid model systems under specified conditions or computer simulations and, therefore, has little relevance to the actual conditions to which polymers are subjected. For the purpose of this book, the "**Denisov-Cycle**" is a good and sufficient model to describe the action of HALS in coatings. Detailed investigations postulating various intermediate species are widely available in the literature [75–82].

The chemical nature of R^1 (Eq. (a) in Figure 3.21) has a significant influence on the formation of >NO. The commercially important HALS carry the following nitrogen substituents: >N-H, >N-alkyl, >N-$(CH_2)_n$-X, >N-$COCH_3$ and >N-O-alkyl (see Table 3.3).

Several experiments have proved that the nitroxyl radical is the active species. It is formed according to Equation (a) and regenerated from an aminoether (Equation (b) and (c) in Figure 3.21). This has been substantiated in an acrylic-melamine clear coat [83, 84], where the generation of peroxide radicals was detected right after the onset of weathering. The concentration of harmful peroxide radicals decreases drastically with formation of the nitroxyl radical. Furthermore, fewer polymeric carbonyl groups, indicating a lower state of degradation, are formed when HALS are added.

Various studies have clearly shown that the nitrogen substituent has a direct influence on the rate at which nitroxyl radicals are formed (Equation (a) in Figure 3.21) [85, 86]. The conversion of >N-R into >NO· takes place much faster if R = CH_3 rather than R = $COCH_3$ (see Figure 3.22).

When equal concentrations (based on piperidine content) of HALS-2 (>N-CH_3) and HALS-5 (>N-$COCH_3$) were mixed into an acrylate/melamine coating and exposed to light, the rate of conversion of HALS-2 into >NO· was much greater than that for HALS-5. The faster formation of nitroxyl radicals with HALS-2 translates directly into a more efficient stabilization of the coating. The results of these tests are discussed in greater detail in Section 4.2.4.

The influence of remote substitution of the pyridine ring of HALS on the formation of >NO· is also evident upon comparison of HALS-5 and HALS-8 (>N-$COCH_3$) [87].

It is evident that nitroxyl radicals play a crucial role in polymer stabilization. It would seem to follow that a homogeneous distribution throughout the coating would be beneficial. However, the degradation of polymer materials requires the presence of oxygen and harmful UV-light. As oxygen uptake is more pronounced near the coating surface than deeper within the coating layers, there is a higher probability of >NO· formation near the surface. The thicker or more dense the coating, the less oxygen can penetrate. Less oxygen leads to lower concentrations of harmful radicals and, in turn, little to no formation of >NO·. Indeed experiments performed in polypropylene and polystyrene [88] have shown that >NO· is only formed where HALS or intermediates of the stabilization cycle (for example >NO-R in Equation (b) in Figure 3.21) are present.

The obvious question arises as to why nitroxyl radicals, which are stable, are not used directly as stabilizers for polymers. The answer lies in their deep orange to red colour [71, 72]. Use of nitroxyl radicals in clear or slightly coloured coatings at concentrations of around 0.5 to 2 %, this would lead to unacceptable discolouration.

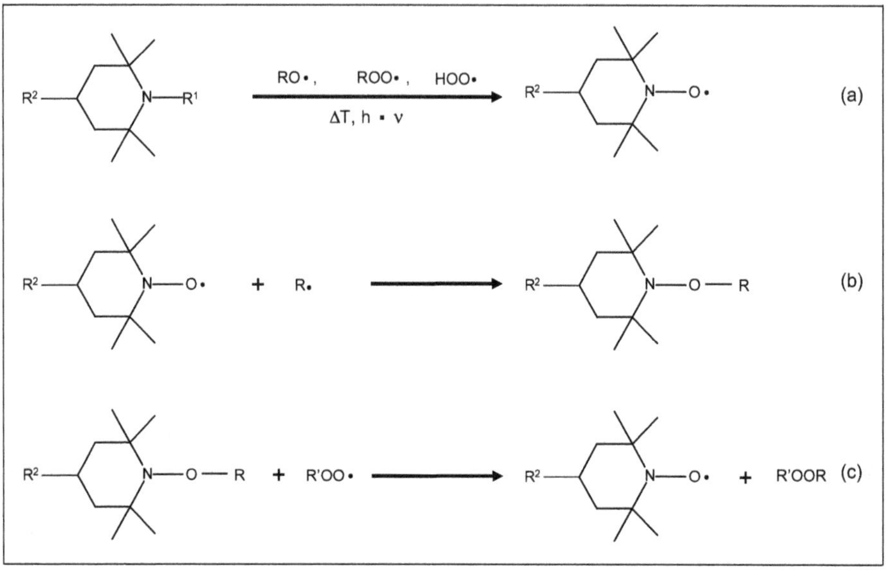

Figure 3.21: The "Denisov cycle" (Equations (b) and (c)) [73]

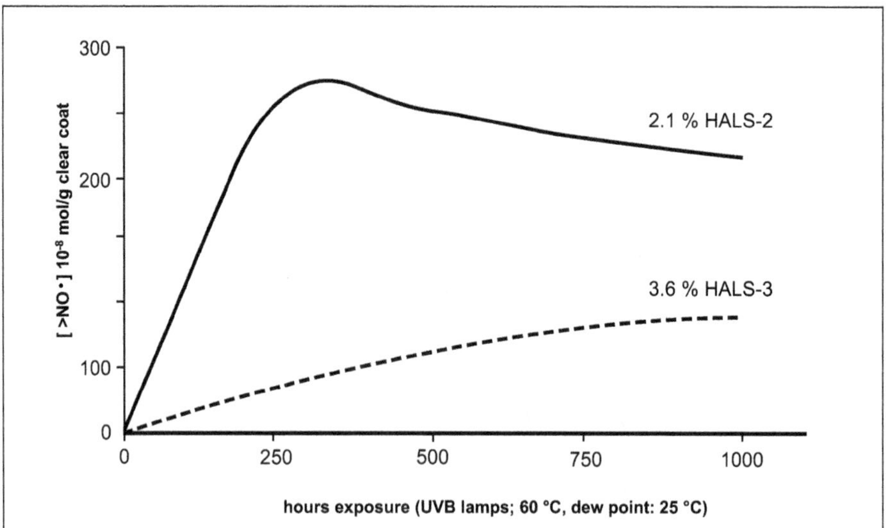

Figure 3.22: Effect of nitrogen substitution on the formation of >NO· [83]
HALS-2: >N-CH_3 HALS-5: >N-$COCH_3$
Identical HALS concentration relative to piperidine nitrogen

As for UV absorbers (see Section 3.2.3.), there are cases where nitroxyl radicals like HALS-12 can be used as stabilizers, for example in wood where a light red colouration is acceptable or even desired (see also see Section 4.3.2.).

3.4 Quenchers

Quenchers are substances that absorb the energy of excited chromophores in the polymer and release it in a non-destructive manner, thus preventing or delaying polymer degradation [89]. Figure 3.23 depicts the quenching process.

The chromophore Ch absorbs energy and is thus promoted to an excited state Ch* (Equation a). In accordance with Equation (b), the quencher Q accepts the energy from chromophore Ch* which reverts into its energetic ground state. Quencher Q*, now in its excited state, can either release this energy in the form of heat (Equation d) or as fluorescent or phosphorescent radiation (Equation c). This reaction sequence can only proceed if the energy acceptor Q possesses excited states that lie lower than that of the excited chromophore Ch* (donor).

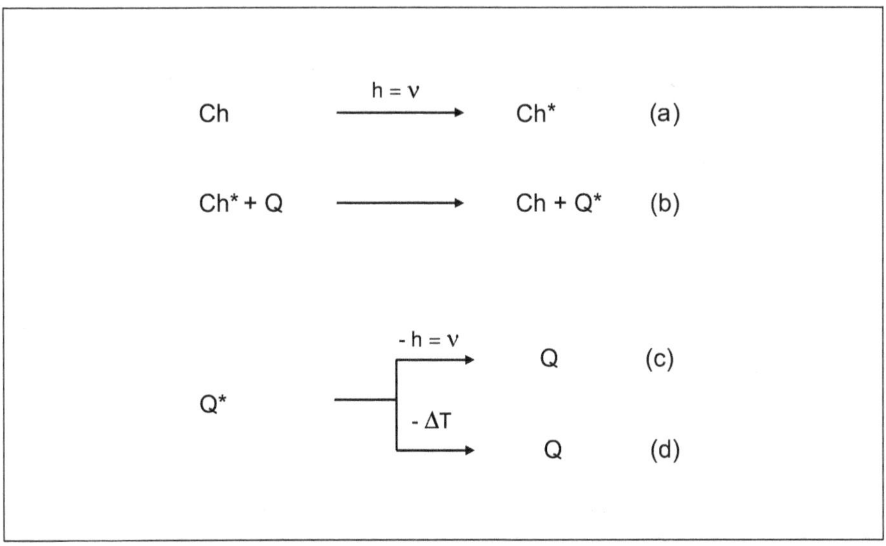

Figure 3.23: Mode of action of quenchers [89]; Ch = chromophore, Q = quencher

There are essentially two modes of energy transfer from the excited chromophore Ch* to the quencher Q [90]:

A) Energy transfer according to the *Förster* cycle [91, 92]

B) Energy transfer through contact, collision or exchange processes.

Option A) can occur over a distance of 5 to 10 nm and requires that the quencher Q to exhibit strong absorption where the chromophore's emits. The *Förster* cycle is also advanced as a possible stabilizing mechanism for UV absorbers with extinction coefficients of ≥10.000 [93] and serves as an explanation why light stabilizers that do not absorb long-wave UV light cannot act as quenchers.

Energy transfer according to B) can take place only if the distance between chromophore and quencher is no bigger than 1.5 nm. This means that energy transfer is directly dependent upon the quencher concentration and the life of the excited chromophore. At the light stabilizer concentrations normally present in coatings, this process is only of secondary importance.

From the practical point of view, quenchers are of interest especially because their effect, like that of free-radical scavengers, is independent of the thickness of the polymeric material. Several nickel compounds are known to be effective quenchers but are unsuitable for coatings because of their inadequate solubility, inherent colour and questionable ecological viability.

3.5 Peroxide decomposing agents

As previously stated, hydroperoxides often play an important role in the photo-oxidative degradation of polymers. In the 1960s, it was hypothesized [58] that certain light stabilizers might break down peroxides. Experimental proof first came in the mid-1970s [94].

In Section 3.3.1 we differentiated between *primary* and **secondary antioxidants**. The latter, typified by thioethers and tri-esters of phosphorous acid (phosphites) [60], can decom-

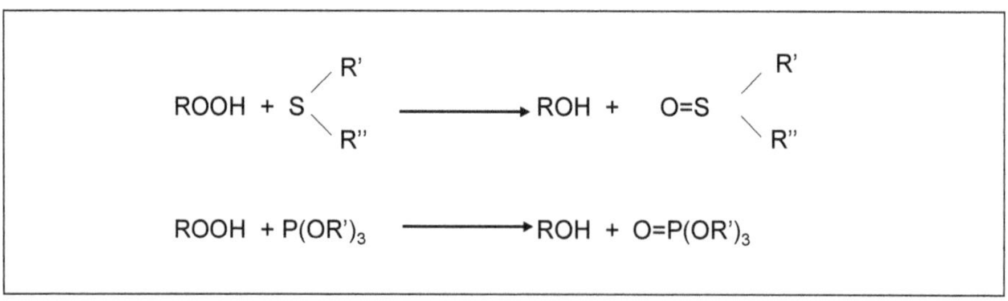

Figure 3.24: Mode of action of secondary antioxidants [60, 95]

pose peroxides through an ionic reaction. This is in stark contrast to *primary antioxidants* which participate in free-radical reactions. Figure 3.24 shows the effect of thioethers and phosphites as peroxide decomposing agents [60, 95].

Secondary antioxidants are most effective when the polymer already contains (hydro) peroxides, for example those formed during the manufacturing process. Other investigations have shown that sterically hindered amines can also act as peroxide decomposing agents [96, 97].

The classes of light stabilizer described in Sections 3.2 to 3.5 are often used in combination. Such light stabilizer blends exhibit synergistic effects in that the effect achieved is greater than the individual components alone. Typical light stabilizer blends include mixtures of

- UV absorbers with sterically hindered amines
- primary antioxidants (phenolic antioxidants) with secondary antioxidants phosphites).

The use of blends of UV absorbers and sterically hindered amines will be discussed later in greater detail.

4 Stabilization of coatings

Coated surfaces must remain intact for long periods if they are to retain their decorative **and** protective effect. It is important to know that the primary purpose of coating a surface, be it metal, plastic or wood, was to protect the substrate, rather than rendering it decorative. Today, however, the two tend to go hand in hand. Many everyday objects such as garden furniture, exterior cladding of buildings, bridges and motor vehicles, are coated with high quality coatings.

Coated surfaces are subject to extreme weather conditions such as UV light, oxygen, moisture and air pollutants, all of which, as mentioned in Section 2, can cause the polymeric material of coatings to deteriorate. Initially, this will affect the appearance of the coated article, suffering a **decrease in gloss** and **change in colour**. Subsequent penetration of moisture into the coating causes **cracking** and **blistering**. Finally, the adhesion of the coating to the substrate weakens and may, in the worst case, lead to complete **delamination**, leaving the substrate completely unprotected. Ultimately, the substrate starts to deteriorate, for example metal surfaces will corrode. This weathering process is shown in Figure 4.1.

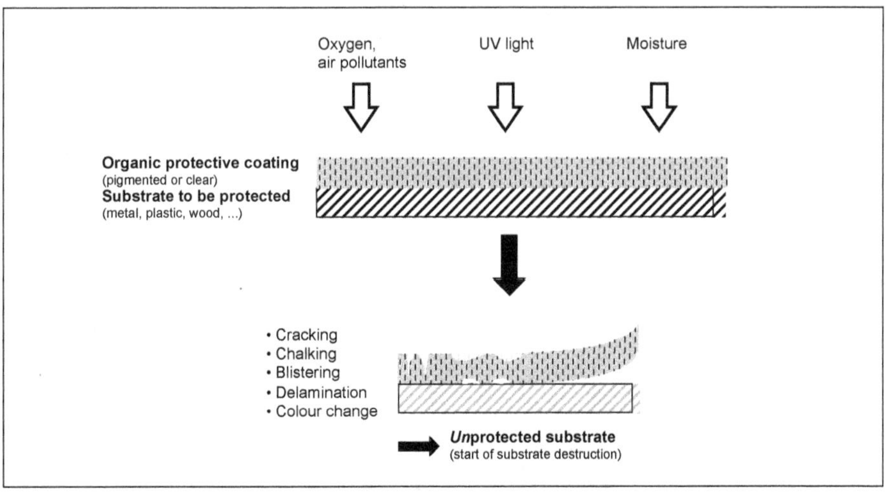

Figure 4.1: Schematic representation of the effects of weathering on coated substrates

Light stabilizers for coatings have been developed to prevent coating deterioration. Here, the emphasis was on automotive coatings, where light stabilization has now become indispensable [20, 22, 35, 98–100]. The importance of this market can easily be seen in the sales figures. In 2014, for example, more than 75 million cars were produced worldwide (see Figure 4.2) [101]. Obviously, the appearance of the vehicle strongly depends upon the long-term stability of the protective coating.

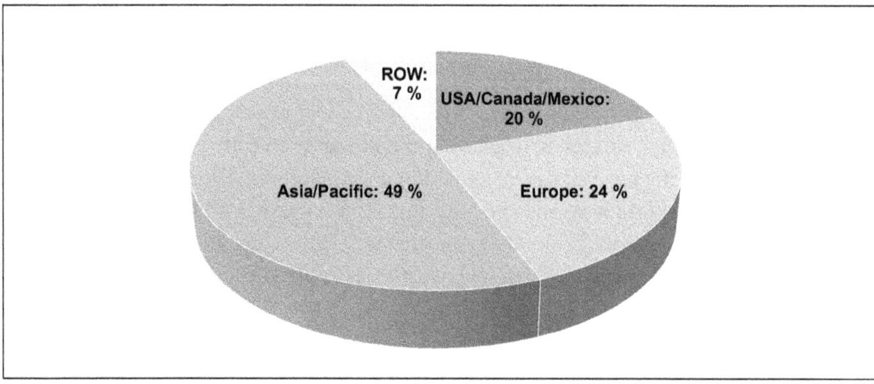

Figure 4.2: Annual world personal vehicle production for 2014 [101]*, worldwide: 75.703.613*

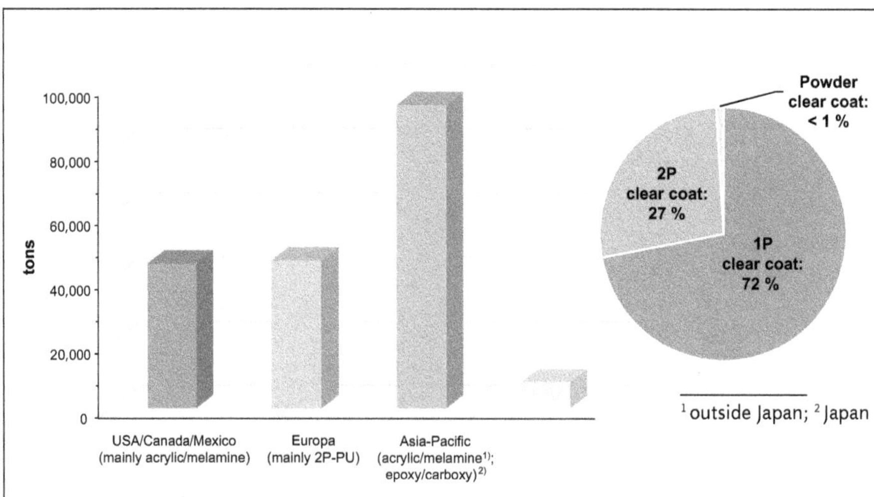

Figure 4.3: Annual world clear coat production (liquid 193.986 tons) for passenger cars and technology split for 2014 [101]

Figure 4.3 shows the worldwide clear coat production in 2014 as well as a split of used technologies.

Assuming that an average of 2.2 kg of **dry** paint per car body are used for the top coating (top 3 layers), which may be pigmented or unpigmented, the car industry alone accounts for the consumption of more than 160.000 tonnes of **dry** paint and all that paint has to be stabilized. Although the emphasis is on the use of automotive coatings in this chapter, the so-called industrial and architectural coatings, for example for tractors, rail cars, plastics and wood, will be discussed in Section 4.3.

4.1 Automotive coatings

There are two types of motor vehicles top coat technologies, namely one-coat (1-coat) and two-coat (2-coat = basecoat/clear coat), the latter thereby being the dominating technology today. Both technologies involve the same kind of surface preparation: phosphating followed by a electrophoretic deposited layer to protect the car body against corrosion.

Subsequently a filler (primer) is usually applied to protect the underlying layer against photo-chemical degradation, even out surface irregularities and provide the necessary stone chip resistance. Such fillers are sometimes coloured to contribute to the targeted colour shade, which is particularly important when pigments with poor hiding power (for example red) are employed. The pigments in the filler may be the same or different from those in the basecoat and may permit a reduction in cost of the first basecoat layer.

After the filler two basecoat layers are applied, the first by electrostatic spraying and the second layer by pneumatic spraying. In two-coat finishes, a clear coat is applied over this pigmented basecoat. In one-coat finishes, the outermost coating is the pigmented top coat itself. Baking conditions for both technologies are 20 to 30 minutes at around 120 °C to 140 °C.

Today there is a clear trend in the automotive industry toward reduction of layers, for example the development of a primer-free processes. Such processes are known as integrated processes meaning that all primer (filler) functions mentioned above have to be integrated in the first basecoat layer.

The coating systems most commonly used for automotive top coats today are:

- Two-pack polyurethane coatings (2P-PU or 2K-PU or 2 component PU are used as abbreviations in the literature; the term 2P-PU will be used throughout this book) which crosslink through reaction of an acrylate polyol and/or polyester polyol with an isocyanate. These systems can also be formulated as high-solids polyurethane coatings (HS-2P-PU). The crosslinking reaction already takes place at room temperature and can be accelerated by using base and/or metal catalysts.
- TSA (thermosetting acrylics) which are crosslinked with a melamine resin. These systems also can also be formulated as high-solids TSA (HS-TSA) which require acid catalysis to support the crosslinking reaction.

- Epoxy resins crosslinked with carboxyl or anhydride resins. Basic and/or metal catalysis may find use depending on the bake temperature.
- Thermosetting alkyd resins, also cured with a melamine resin. These binders are used mainly for one-coat finishes.

In the 1990s, the development of powder clear coat gained major interest. Such clear coats were based primarily on carboxyl and anhydride-functional acrylic resins cured with epoxy resins. Today, this technology is of minor importance.

The crosslinking reactions of the aforementioned systems are the subject of several publications [102–108]. In addition, there are several other options for high temperature or low to room temperature curing [109–111].

Most of the coatings used today still contain organic solvents. Water-based and solvent-free systems are also available or in development, such as powder coatings but also radiation curing coatings [112].

Many studies have been published on the degradation mechanism of coatings, including those based on acrylic/melamine (TSA) and hydroxyl-functional acrylates and polyester/isocyanate (2P-PU) [113–118]. There is, therefore, no need to describe these process in detail. Suffice to say that each weathering parameter, UV light, moisture, temperature and air pollutants, can alone cause polymer degradation but the combination of all these parameters causes the greatest damage to the polymer.

The effect of air pollutants such as acid rain on polymer degradation has also been subject of several studies [119, 120]. It is important to note that the photo-oxidative degradation of polymers by sulphur and nitrogen oxides can be greatly reduced by use of suitable light stabilizers [121].

4.2 Light stabilization of automotive coatings

In Section 3, the options for stabilizing polymers were discussed in detail. Two of these are especially relevant for coatings, especially automotive coatings [21–24, 97, 122–124]:

a. **UV absorbers** to compete in the absorption of UV light in the wavelength range of 290 to 380 nm, in combination with
b. sterically hindered amines (HALS = hindered amine light stabilizers) that trap the radicals formed during polymer degradation.

The following Chapters 4.2.1 to 4.2.3 will focus on UV absorbers and HALS used in automotive coatings and the special requirements they need to fulfil. Exposure results will be discussed in Chapter 4.2.4

4.2.1 Two-coat systems

UV absorbers are incorporated in automotive clear coats to prevent pigment discolouration and photo-chemical degradation of the binders caused by photo-chemical reactions. These photo-chemical degradation processes can lead to loss of adhesion and complete delamination of the clear coat coat from the basecoat or, in the worst case, of the entire coating system from the surface.

In two-coat systems, the type of pigment used for the basecoat, for example metallic, coloured or white pigment, has a direct influence upon the weathering resistance of the clear coat [125]. Figure 4.4 shows the optical conditions in two-coat finishes [126].
A small part of the incident light is reflected by the coating surface. The extent of reflection will depend on the angle of incidence as well as on the refractive index of the clear coat. Light that passes through the clear coat is absorbed or reflected by the basecoat. The light reflected by the basecoat again passes through the clear coat where it is either absorbed or reflected, depending on the angle of incidence. Thus, the clear coat is exposed to a high intensity of potentially harmful UV light.

Clear coats on metallic finishes are subjected to even harsher conditions since metallic pigments can both scatter and reflect light. Addition of a UV absorber to the clear coat changes the composition as well as the intensity of the transmitted and reflected light, depending on its concentration and absorption characteristics.

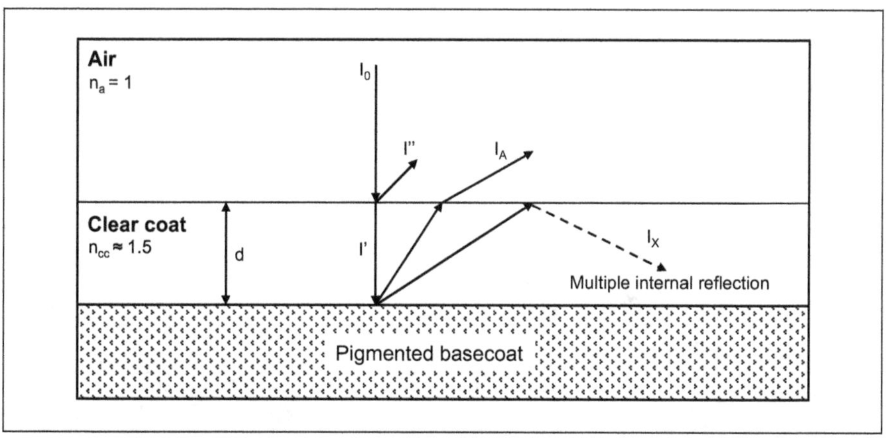

Figure 4.4: Schematic representation of optical conditions in 2-coat finishes [126]

n_a: refractive index air | *n_{cc}: refractive index clear coat*
d: thickness of clear coat | *I_0: incoming light*
I'': light reflected at surface | *I': light passing through clear coat*
I_A: light reflected at basecoat | *I_x: internally reflected light*

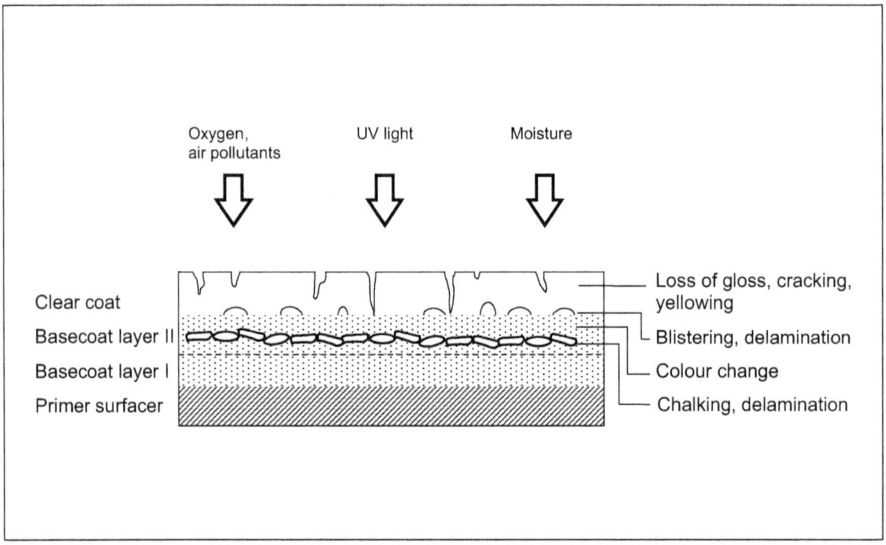

Figure 4.5: Deterioration of basecoat/clear coat systems caused by weathering

Figure 4.6: Cracking in a two-coat metallic coating

The colour of the pigments used in the basecoat also influence the heat absorption of the coating [127] and any increase in temperature will accelerate reactions leading to degradation. Depending on the latitude, dark, horizontal surfaces can reach temperatures of up to 100 °C or more when exposed to intensive sunlight.

As mentioned in Chapter 3.3, sterically hindered amines (HALS), which do not absorb at wavelengths above 250 nm, trap radicals and are mainly responsible for gloss retention and prevention of crack formation in coatings. Figures 4.5 and 4.6. show the weathering of an unstabilized two-coat system and a metallic coating.

Recommended UV absorber concentrations in clear coats range from 1.5 to 3 % based on the solid binder content, for a dry film thickness of about 40 µm. For thinner film the UV absorber concentration will have to be increased in accordance with Lambert-Beer's law (see Chapter 3.2.1) in order to achieve the same stabilizing filter effect.

The HALS concentration should be around 0.5 to 2 %, calculated on the solid binder content. Unlike UV absorbers, the stabilizing effect of HALS is largely independent of film thickness.

In general, sufficient stabilization of clear coats can be achieved by a blend consisting of about 1.5 to 2 % UV absorber and 1 % HALS.

Depending on the binder and pigment used, weathering resistance of the basecoat can be further improved by adding more stabilizer. The stabilization of basecoats is discussed in greater detail in Chapter 4.2.4.6.

Besides UV absorbers and HALS, it is also possible to use phenolic antioxidants and/or phosphites (peroxide decomposing agents) in automotive finishes. In contrast to the former that provide long-term protection of the coating, phenolic antioxidants and phosphites are employed for the protection of the binder during the high temperature baking process. Typical concentrations range between 0.05 and 0.5 %, calculated on the solid binder.

The weathering of two-coat finishes is discussed in Chapter 4.2.4. The stabilization of one-coat finishes is slightly different and treated separately Chapter 4.2.5.

4.2.2 Specific requirements of UV absorbers in coatings

UV absorbers used in coatings must meet certain requirements [16], which can be divided into **primary** and **secondary requirements**, as detailed below.

I) Primary requirements

- high extinction coefficient (great effectiveness)
- broad spectral coverage between 290 and 380 nm
- steep absorption edge
- photo-chemical stability

Extinction coefficients, absorption bands and the slope of the absorption edge were discussed already in Chapter 3.2.3., (see also Table 3.2 and Figures 3.5, 3.4 and 3.12). The photo-chemical stability of different UV absorbers will be discussed in Chapter 4.2.6.1.

II) Secondary requirements

- excellent solubility in the solvents commonly used in the coating industry; ideally a liquid
- ease of incorporation in water-borne systems, if possible without having to use co-solvents
- ease of incorporation in solvent-free systems such as powder coatings; the melting point of the UV absorber should be in the range of standard extrusion temperatures
- radiation-curing coatings. The UV absorber should be liquid if possible
- heat resistance (low volatility at 80 to 150 °C)
- no unwelcome side reactions or interaction with other components of the formulation such as catalysts
- good compatibility with polymers (no exudation from crosslinked polymers)
- resistance to extraction

Two comments should be made about the above secondary properties:

1. there is no single product which meets the requirements for every possible application
2. this means that all UV absorbers represent the best compromise possible.

Now, the most important secondary properties will be treated in more detail.

4.2.2.1 Solubility and compatibility of UV absorbers

UV absorbers must be soluble in all kinds of different formulations, as well as in solvents typically used in the coating industry. In view of the increasing limitations regarding the use of organic solvents (VOC regulations, classification as air pollutants), it is essential that additives such as UV absorbers can be incorporated into the coating formulations without aid of co-solvents. This is possible if the additive is readily soluble in one of the solvents used in the formulation. Ideally, the additive can be stirred directly into the liquid formulation. Barring that, a small amount of the solvent can be used to make an additive solution, as long as the viscosity of the formulation is not greatly affected.

Ideally, UV absorbers should be liquid, although they may contain small amounts of non-toxic solvents for handling reasons. Table 4.1 gives an overview of the solubility characteristics of common UV absorbers.

Liquid products are miscible in standard formulations, including high-solid formulations, and the use of additional solvent is not necessary (solubility >50 %). Certain solid UV absorbers, on the other hand, show very poor solubility, for example BTZ- 2, BTZ-3 and oxanilide-1, and can only be incorporated into the formulation with the help of a co-

Table 4.1: Solubility of UV absorbers in different solvents

UV absorber	**Solubility (% w/w) @ 20 °C**			
	Methyl ethyl ketone	n-Butanol	Butyl acetate	Xylene
BTZ-3	5.5	0.3	4.5	12
BTZ-4	n.a.	17	> 30	> 50
BTZ-5	13	n.a.	n.a.	11
BTZ-6	30	n.a.	28	> 50
BTZ-7	> 50	> 50	> 50	> 50
BZT-8	> 50	> 30	> 30	> 50
HPT-1	> 50	> 50	> 50	> 50
HPT-2	n.a.	9	12	20
HPT-4	n.a.	< 1	17	> 50
HPT-5	5.7	< 1	4.3	n.a.
HPT-6	> 50	n.a.	> 50	> 50
Oxanilide-2	> 50	> 50	> 50	> 50

solvent. UV absorbers with poor solubility can also cause problems in the cured film if they are incompatible with or insoluble in the binders. This can result in re-crystallization of the UV absorber leading to turbid coating or even to exudation resulting in a deposit formation on the surface.

In the case of solvent-free systems, one must differentiate between liquid and powder coatings.

Typical examples of the former are radiation curing coatings where clear advantages exist for liquid UV absorbers.

In contrast, solid UV absorbers are preferred for powder coatings due to easier handling and better storage stability. Here, liquid additives would lower the glass transition temperature of the coating. However, the UV absorbers must dissolve in the melted polymer during extrusion. If the melting point of the UV absorber is higher than the extrusion temperature, inhomogeneous distribution in the powder coating can result potentially leading to surface defects due to crystalline deposits of the UV absorber. Optimally, the melting point of the UV absorber should match the extrusion temperatures used in the production of powder coatings. In general, melting points of around 70 to 100 °C will ensure that there are no compatibility problems.

4.2.2.1.1 Water-borne systems

Most of the light stabilizers initially developed for use in solvent-based coatings can also be used in water-based systems. The "only" challenge is their incorporation in aqueous systems. Larger coating or resin manufacturers may be able to incorporate the light stabilizers in the organic phase, for example during resin synthesis. For most coating suppliers, however, this is not possible so efforts have been directed to the development of water compatible product forms that can be easily incorporated in the final water-based formulation.

The following list gives an overview of product forms developed for water-borne applications [128]:

- **hydrophilic** light stabilizers such as the BTZ-7. The polarity of the side chain introduced is sufficiently high to permit its incorporation in most formulations without resorting to the use of co-solvent.
- **solid state dispersions**: solid UV absorbers and/or sterically hindered amines that have been dispersed in water with the help of dispersants and small amounts of organic solvent (for example glycol ether). The solids content (active substance) is generally around 50 %. Despite a certain risk of sedimentation, these products are readily re-dispersed under stirring. Their usefulness in ambient curing systems is sometimes limited due to insufficient solubility in the polymer matrix.
- **emulsions**: light stabilizers that are converted into a water compatible form by emulsification with non-ionic emulsifiers. Due to high levels of emulsifiers used, the water sensitivity of the final coating may be increased.

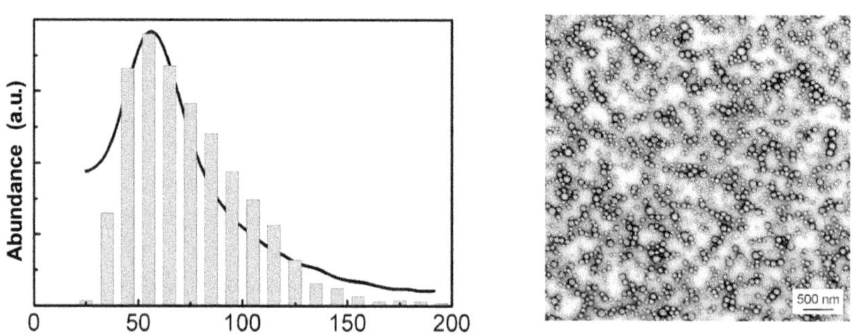

Figure 4.7: Particle size of nano dispersions

Table 4.2: Examples for water compatible UV absorbers

UV absorber	Technology	Chemical composition*
BTZ-9	hydrophilic	BTZ-7
BTZ-10	nano dispersion	BTZ-8
BTZ-11	solid state dispersion	BTZ-3
BTZ-12	solid state dispersion	BTZ-2
BTZ-13	solid state dispersion	BTZ-5
HPT-7	nano dispersion	HPT-1
HPT-8	solid state dispersion	HPT-3
HPT-9	nano dispersion	HPT-4
HPT-10	nano dispersion	HPT-6
BP-4	solid state dispersion	BP-1

* see also Figs. 3.8a, 3.8.b, 3.9.a, 3.9b and 3.10

- **Neat (Novel Encapsulated Additive Technology)** [129]: relatively new on the market, light stabilizers are dissolved, rather than "encapsulated", in an acrylic type polymer matrix characterized by an average particle size < 250 nm (Figure 4.7). Such nano-dispersions are VOC-free and, due to their small particle size, can be also used in ambient curing systems providing highly transparent coatings. The solids content varies depending on the individual product and is typically 40 to 50 %, half to two thirds of the weight is active substance (UV absorber) and the remaining polymer. These dispersions can be used in any loading that the application requires.

Examples for water compatible UV absorbers are shown in Table 4.2.

Water-soluble light stabilizers for cosmetic applications are not suitable for water-based coatings. Typically, they cause destabilization or even coagulation of the formulations or a massive increase of the water sensitivity of the cured film.

4.2.2.2 Volatility of UV absorbers

The UV absorbers used in automotive coatings must resist high temperatures during their application and the lifetime of the vehicle. Temperatures under the bonnet and elsewhere in the car can exceed 100 °C, especially in cars with darker coatings and/or in the tropical regions of the world. In car production, problems can arise so that car bodies remain longer in the baking oven than the stipulated time of about 30 minutes. UV absorbers should therefore not only exhibit high thermal stability but also low volatility.

Table 4.3: Volatility of UV absorbers in an acrylate/melamine clear coat after bake and overbake; DFT: 40 µm

UV absorber (2%)	% loss after bake (140 °C, 30')	% loss after overbake (160 °C, 45')
BTZ-2	39.0	89.0
BTZ-4	2.6	21.3
BTZ-7	0.4	3.0
BTZ-8	1.7	16.4
HPT-1	0.7	0.7
HPT-2	0.5	6.4
HPT-4	0.1	0.1

Table 4.4: Volatility of UV absorbers in two different 2P-PU clear coats. bake: 30 min at 130 °C; DFT: 20 µm

UV absorber (2 %)	% loss after overbaking (150 °C, 60')	
	clear coat I	clear coat II
BTZ-2	> 60	> 90
BTZ-3	0 to 1	1 to 2
BTZ-8	2 to 3	5 to 7
HPT-1	0	0
BP-2	> 30	> 40
Oxanilide-1	> 30	> 50
Oxanilide-2	5 to 7	6 to 9

Various methods of reducing the volatility of UV absorbers have been described in the literature [130]. Two of these stand out:

a. increasing the molecular weight of the UV absorber
b. chemically binding the UV absorber to the polymer framework (during the crosslinking reaction or during binder synthesis).

Both methods have their limitations however because of

- insufficient solubility or compatibility of oligomeric or polymeric UV absorbers
- lower "formulation latitude" or change of binder properties, if UV absorbers are built into the binder itself.

The specific benefits of using UV absorbers which are able to react with the binder system during the crosslinking reaction will be discussed in a separate Chapter.

Tables 4.3 and 4.4 list the heat resistance of some UV absorbers. The results do not only show the obvious dependence of the volatility on the temperature but also the variation with both film build and polymer type (for example varying if cross densities). Further influences include paint constituents which can react with hydroxyl groups of the UV absorber (though not with the phenolic hydroxyl groups needed for stabilization).

Table 4.3. shows the results in an acrylate/melamine clear coat (DFT: 40 µm) after bake and subsequent overbake. The results clearly show that the volatility is significantly more pronounced for small(er) UV absorber molecules.

In Table 4.4. volatility data in different 2P-PU clear coats applied at a dry film thickness of 20 µm are listed. Note that volatility can also vary with crosslink density.

The figures clearly show that, BTZ-2, BP-2 and oxanilide-1 are unsuitable for applications involving very high temperatures. Taking the inevitable experimental errors into account, the volatility of a UV absorber should not be greater than 5 %.

4.2.2.3 Reactable UV absorbers

The benefit of UV absorbers bearing functionalities that can react with binder components are described in Figures 4.8 to 4.10 [131, 132].

Two different UV absorbers were investigated. BTZ-4 as an example of a non-reactable and HPT-1 as an example for a reactable UV absorber (secondary OH in side chain, see Figure 3.9a in Chapter 3.2.3).

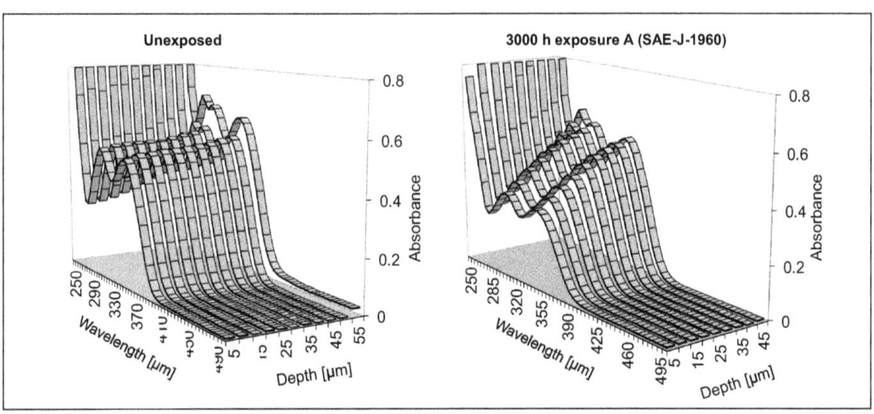

Figure 4.8: UV microscopy profile: 2P-PU clear coat over metallic basecoat containing BZT-4 in basecoat and clear coat, substrate: steel

The UV absorbers were incorporated in the same amount in both in the 2P-PU clear coat and in the basecoat. In one case, the clear coat/basecoat system was applied on a steel substrate and in the other case over a thermoplastic polyolefin (TPO), a widely used plastic substrate in the automotive industry. The UV absorber content was determined by ab-

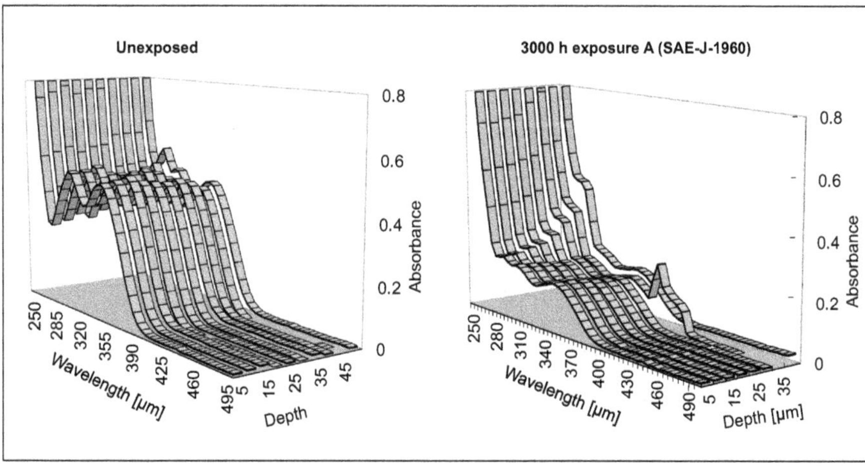

Figure 4.9: UV microscopy profile: 2P-PU clear coat over metallic basecoat containing BZT-4 in basecoat and clear coat; substrate: thermoplastic polyolefin (TPO)

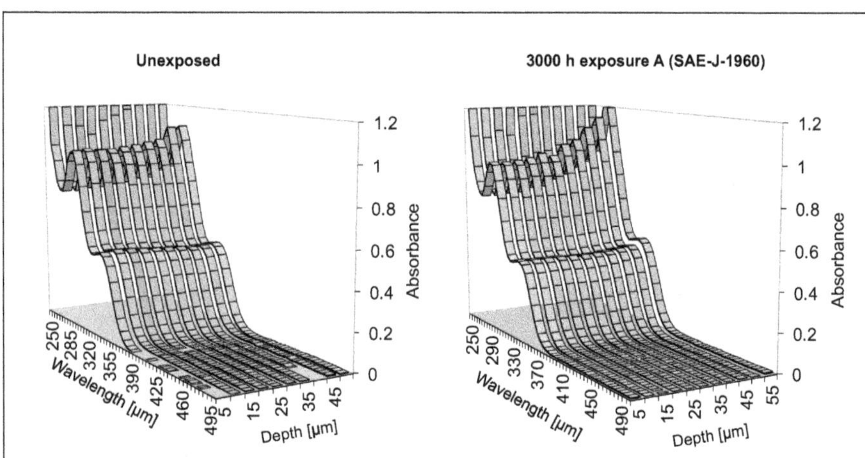

Figure 4.10: UV microscopy profile: 2P-PU clear coat over a metallic basecoat containing HPT-1 in basecoat and clear coat, substrate: thermoplastic polyolefin (TPO)

sorbance as a function of the layer depth using microtomed coating slices. Spectra were recorded before and after accelerated weathering.

As expected the unexposed coating in Figure 4.8 shows a more or less homogeneous distribution of BTZ-4 throughout the coating. After 3000 h exposure (details see Chapter 4.2.4.1) there is only a minor loss of the UV absorber in the top layers but there is no loss of UV protection in the deeper layers down to 50 μm.

The same coating system applied over TPO (Figure 4.9) presents a completely different picture. Whereas the unexposed sample exhibits more or less the same homogeneous distribution of BTZ-4 compared to steel substrate, the exposed one shows a dramatic loss of BTZ-4 all the way down to 50 μm, indicating insufficient UV protection. On TPO BTZ-4 behaves completely different compared to the application over a steel substrate!

Now, let us have a look at the behaviour of the reactable HPT-1. For the steel substrate, the situation is exactly the same as in case of BTZ-4, a homogeneous distribution before and after exposure.

But now even the TPO substrate (Figure 4.10) shows only a small loss in the very top layers but no loss at all in the deeper layers, a significant difference compared to the non-reactable BTZ-4. The results can be explained as follows:

- The (slight) loss of UV absorber over steel in the very top layers by migration to the surface and loss due to erosion

and/or

- photo-chemical degradation, which implies that HPT-1 is more photostable compared to BTZ-4 (details concerning photostability of UV absorbers see Chapter 4.4.1)

The significant loss of BTZ-4 compared to HPT-1 on the TPO substrate clearly indicates significant migration into the plastic substrate the reactable HPT-1 providing complete UV protection even after 3000 hours of exposure! It is therefore highly recommended to use reactable UVAs in coatings applied over plastic substrates. The same is valid for HALS which are discussed in Chapter 4.2.3.3.

4.2.2.4 Effect of UV absorbers on coating colour

UV absorber chromophores are usually yellowish (tailing absorption into the near visible), the colour intensity thereby depending on whether the substance is a liquid or a solid. If UV absorbers are observed in a test tube, the liquids will appear more coloured than solids. This is due to the fact that one can look right through the liquids to the other side of the test tube, whereas one sees only the solid that is in immediate contact with the test tube wall. Thus, the colour of the UV absorber itself can only give an approximate indication of the colour it will impart to the cured coating.

Although the colour of the wet paint (in-can colour) is regarded important or even decisive in refinish applications, the colour of the cured coating is more realistic. Some practical

Table 4.5: Effect of UV absorbers on clear coat yellowing

UV absorber (1.5 %)	Yellowness index b*		
	40 µm	60 µm	40 µm + overbaking 150 °C, 1 h
Without	- 3.8	- 3.7	- 2.8
Oxanilide-2	- 3.5	- 3.3	- 2.7
BTZ-1	- 3.4	- 3.1	- 3.2
BTZ-4	- 2.1	- 1.6	- 1.6
BTZ-6	13.8	16.6	14.7
BTZ-8	- 2.5	- 2.1	- 1.7
HPT-1	- 2.9	- 2.5	- 2.0
HPT-4	- 1.8	- 0.7	- 2.0
HPT-5	- 1.3	- 0.4	- 1.0
HPT-6	- 0.8	0.3	- 0.6
HPT-1/HPT-4 (2/1)	- 2.5	- 1.9	0.5
BP-2	- 2.8	- 2.4	- 1.7

clear coat: 2P-PU DFT: 40 µm and 60 µm
basecoat: white
bake: 130 °C, 30' overbake (40 µm only): 150 °C, 1 h

test results are shown in Table 4.5. The evaluation was made with a 2P-PU clear coat, containing different UV absorbers (1.5 % on solid binder) at different film thicknesses on a white basecoat. After baking and overbaking (only the 40 µm thick coating) the yellowness index b* was measured (a negative yellowness index means a shift on the yellow/blue axis into the blue).

Independent on the UV absorber class, it was found that the colour of the clear coat intensifies with increasing dry film thickness. The same effect is observed when increasing the UV absorber concentration by keeping the dry film thickness. Oxanilide shows the least increase of b*, which is in line with the blue shift in its absorption spectrum shown in Figure 3.4. The more the UV spectrum is shifted toward the visible (380 to 400 nm and above), the higher is the initial yellowness index: HPT-2 is typified by a lower b* compared to HPT-4, HPT-5, and HPT-6 (see Figure 3.13 for the absorption spectra).

4.2.2.5 Unwelcome side reactions

Some UV absorbers can react or interact with other formulation constituents to cause unwelcome side effects such as inadequate curing or colour deviations.

Table 4.6: Yellowing of a 2P-epoxy clear coat due to benzophenones and benzotriazoles'

UV absorber (1.5 %)	**Yellowness index YI**	
	after curing	**+ 48 h at RT**
BTZ-3	- 0.3	13.0
BTZ-7	- 0.4	6.5
BP-3	- 0.4	7.1

clear coat: amino functional/epoxy functional acrylate DFT: 30 µm
basecoat: white bake: 60 °C, 30'

Phenolic UV absorbers such as benzophenones and benzotriazoles (see Figures 3.9 and 3.11) tend to interact with strong bases, for example primary amines, metal-based catalysts such as zinc octoate and aluminium tris-acetylacetonate as well as with other basic formulation components such as aminofunctional acrylates [133, 134]. In case of bases, these interactions lead to a deprotonation of the phenolic hydroxyl group of the UV absorber resulting in a red-shift of its absorption and the generation of colour. Certain metals can perturb the intra-molecular hydrogen bridge and lead to the formation of quinoidal structures having a characteristic intense yellow colour. Table 4.6 illustrates this strong yellowing in a 2P-epoxy clear coat.

No yellowing is evident immediately after curing. After ageing for 48 hours at room temperature, however, all three UV absorbers produced a distinct yellow discolouration.

In contrast to benzophenones and benzotriazoles, hydroxyphenyl-s-triazines (mono-resorcinyl-triazines HPT-1 to HPT-4), the third class of phenolic UV absorbers, due not interact with other coating components to produce undesired colour [37, 135]. This recommends their use for the stabilization of coatings that are basic or cured with metal catalysts (see Figures 4.11 and 4.12).

Figure 4.11 shows the behaviour of three different phenolic UV absorber in a 2P-epoxy clear coat based on an amino and epoxy-functional acrylate with a dry film thickness of 30 µm. Although no yellowing was observed after 24 hours, films containing BP-3, BTZ-3 and BTZ-4 showed very strong yellowing (high yellowness index YI) after 48 and 72 hours. On the other hand, the film stabilized with HPT-1 remained virtually unchanged over the same time period.

Similar results are obtained with clear coats cured with metal catalysts. A typical example is a binder system consisting of a copolymer based on vinyl monomers containing oxirane and alkoxysilane [136, 137]. The crosslinking of such a clear coat formulation is catalysed with an aluminium chelate catalyst.

The undesirable metal interaction of UV absorbers can be easily illustrated by a model experiment. 2 % aluminium tris-acetylacetonate and UV absorber are added to a standard

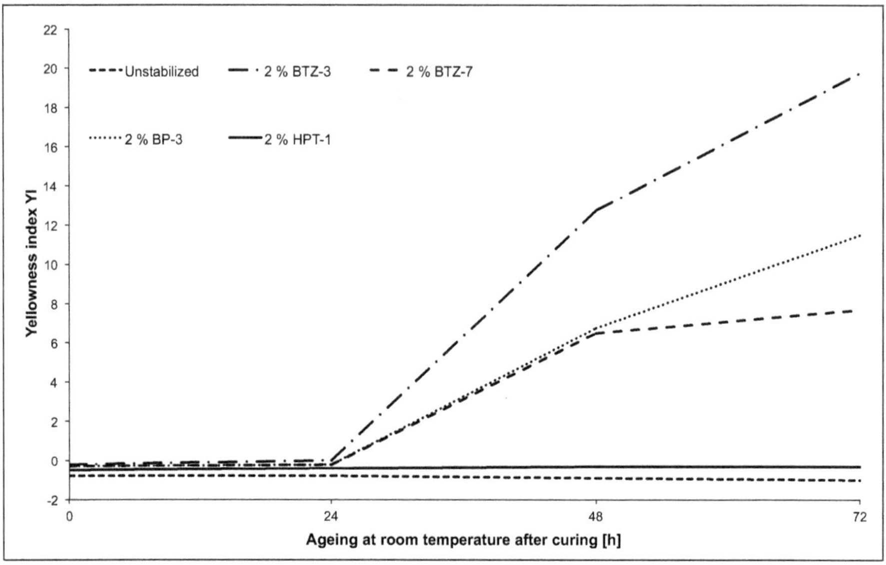

Figure 4.11: Yellowness index YI of a 2P-Epoxy clear coat containing various phenolic UV absorbers

clear coat: aminofunctional/epoxyfunctional acrylate; 30 µm, basecoat: white, curing: 30 min at 60 °C

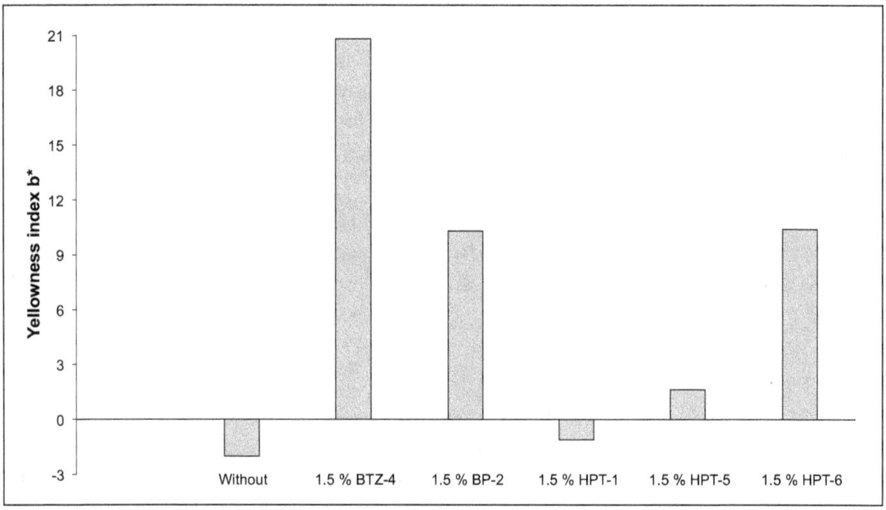

Figure 4.12: Model experiment showing the effect of UV absorbers on the yellowing of metal catalyzed clear coats

clear coat: 2P-PU, catalyst: 2 % aluminium tris acetylacetonate, DFT: 40 µm, bake: 130 °C, 30′

2P-PU clear coat. The formulation was applied to a white basecoat and baked for 30 minutes at 130 °C. The dry film thickness was 40 µm. Figure 4.12 shows the dependence of the film's yellowness index b^* on the UV absorber.

Again, there is strong yellowing in the case of BP-2 and BTZ-4 ($b^* >> 10$), whereas HPT-1 provides results comparable to the clear coat without UV absorber. The number of OH-groups at the triazine ring increases from HPT-1 to HPT-5 and HPT-6. As discussed earlier this leads to a red-shift of the absorption spectrum (Figure 3.13) but also to a more pronounced tendency to interact with a metal catalyst. It also changes the behaviour of these compounds in a metal catalysed system. HPT-5 and HPT-6 show significant higher yellowing b^* than that containing HPT-1. The experimental results depicted in Table 4.5 (DFT: 40 µm) are directly comparable with those of Figure 4.12; the coating system and film thicknesses are the same, the only difference is the presence of the metal-catalyst in the latter. These results clearly show that knowledge of the composition of the coating formulation is indispensable for the selection the right UV absorber.

4.2.2.6 UV absorbers and photoinitiators

UV-curable coating formulations contain so-called "photoinitiators", the substances that initiate the crosslinking reaction. Of these, the radical photoinitiators find greatest use. Photoinitiators absorb mainly in the UV region between 300 and 400 nm, exactly where UV absorbers should absorb the harmful light. Therefore, upon first glance, it would not seem possible to perform UV curing in the presence of UV absorbers or to stabilize UV-curable coatings in the same manner as thermally curing coatings. This apparent contradiction might be responsible for the fact that such systems are not as widely used as they deserve to be. Correctly formulated UV-curing coatings can also enjoy perfect weatherfastness. Nonetheless, until now, UV-curing coatings have been used mainly for interior applications.

Figure 4.13 shows four examples of radical photoinitiators: PI-I and PI-II belong to the class of α-hydroxyketones (α-cleavage photoinitiators). PI-III and PI-IV are a bis-acyl phosphine oxide type photoinitiators. These compounds can be used alone or as mixtures. Such blends of PI-I or PI-II with PI-III or PI-IV type photoinitiators produce very interesting results, especially in pigmented systems, where light absorbing pigments compete with the photoinitiator for the available light [138]. Here, the special synergy of these PI blend is evident: while α-hydroxyketones improve surface hardness, the bis-acyl phosphine oxides provide full through cure.

Given the efficient curing of pigmented systems, it Is not surprising that various studies have shown that UV absorbers can also be used together with photoinitiators [50, 139–145]. Here it is important to realise that the most light absorbed by the photoinitiator must not necessarily be at its absorption maximum [146]. Such coatings will cure properly if the UV absorber allows penetration of a sufficient quantity of UV light in the deeper layers of the

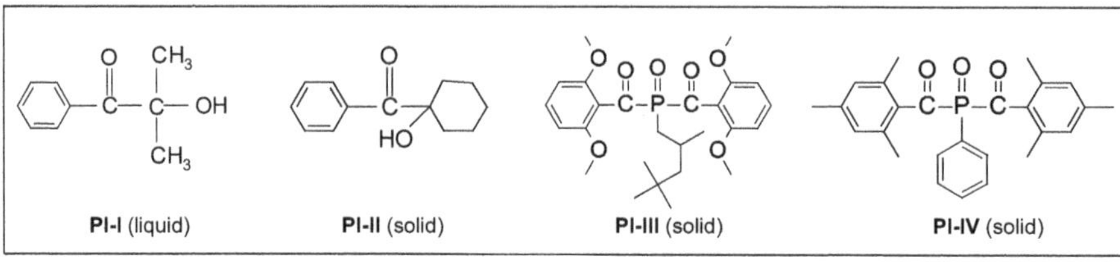

Figure 4.13 Examples of photoinitiators (PI) for UV-curable coatings

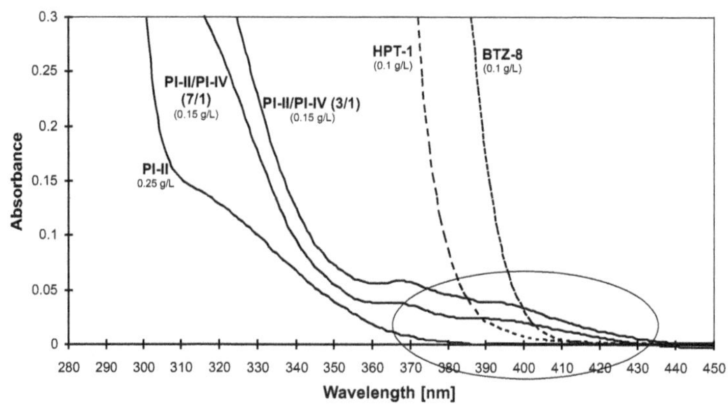

Figure 4.14: UV spectra of Examples of PI II/PI IV combinations compared to BTZ-8 and HPT-1, solvent: $CHCl_3$; 1 cm cell

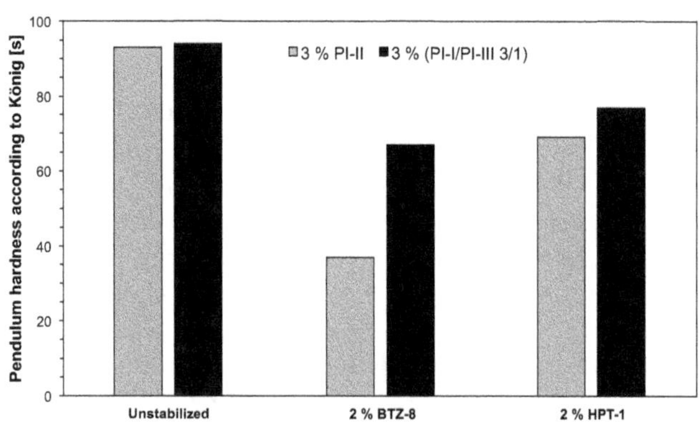

Figure 4.15: Effect of UV absorber and photoinitiator on the curing of a UV-curing clear coat
clear coat: aliphatic epoxy acrylate/aliphatic urethane acrylate/tripropylene glycol acrylate; 40 µm, basecoat: silver metallic (pre-baked), curing: 2 x 80 W/cm, 2 x 10 m/min

Table 4.7: Effect of UV absorber and photoinitiator on the pendulum hardness of a UV-curing clear coat

Clear coat applied on top of various basecoats	**Pendulum hardness [s] after curing**	
	without UV absorber	**1.5 % BTZ-8**
1 % PI-II: white basecoat	100	76
3 % PI-II: white basecoat	189	179
3 % PI-II: dark blue metallic basecoat	95	76
5 % PI-II: dark blue metallic basecoat	95	83

clear coat: aliphatic polyester triurethane triacrylate/hexanediol diacrylate (60/40); 40 µm
basecoat: a) white b) water-borne, dark blue metallic (pre-baked)
curing: 2 x 80 W/cm, 2 x 10 m/min

coating as shown in Figure 4.14. For this study, a UV absorber/photoinitiator ratio of 2/3 was employed which corresponds to a clear coat stabilized with 2 % UVA and cured with 3 % photoinitiator. The spectra of combinations of PI-II with PI-IV reveal moderate absorption up to ca. 430 nm, indicating that sufficient UV light is available to cure the clear coat [147]. HPT-1 offers even more benefits since its absorption cut-off is at lower wavelengths than BTZ-8. Thus, more UV light can be absorbed by the photoinitiator and better through curing is achieved (see Figure 4.15).

Tables 4.7 and 4.8, as well as Figure 4.15 show the results obtained UV-cured clear coats containing UV absorbers applied on different basecoats. Table 4.7 shows that adequate UV curing with a given light source depends on the photoinitiator concentration and the colour of the substrate being coated (in this case the basecoat).

If the PI-II concentration is sufficiently high, the pendulum hardness of the clear coat stabilized with UV absorber will be only slightly lower than that of the unstabilized sample.

Figure 4.15 shows the effect of the type of photoinitiator and UV absorber on the curing characteristics of a UV-cured clear coat. If PI-II is used together with BTZ-8, curing is distinctly worse than with the combination PI II/HPT-1. If one replaces PI II by a combination of P-II and PI-III, an increase of the pendulum hardness is observed. With the UV absorber HPT-1 one observes an improvement in curing behaviour as well with the P-II and PI-III photoinitiator combination but this is less pronounced as with BTZ-8. As discussed above, the slightly lower absorption of HPT-1 at higher wavelengths leaves more UV light for the photoinitiator resulting in more efficient curing the clear coat.

Table 4.8: Effect of UV absorber and photoinitiator on the yellowing of a UV curable

Photoinitiator (3%)	**UV absorber (2%)**	**Yellowness index ... minutes after curing**	
		15	**60**
PI II	-	6.5	5.6
	BTZ-8	11.2	8.9
	HPT-1	2.7	1.5
PI-I/PI III (95/5)	-	6.7	5.7
	BTZ-8	14.3	10.9
	HPT-1	3.3	2.2
PI-II/PI III (3/1)	-	7.8	6.9
	BTZ-8	13.6	11.5
	HPT-1	5.8	4.9

clear coat: aliphatic polyester triurethane triacrylate/hexanediol diacrylate (60/40); 40 µm
basecoat: white
curing: 2 x 80 W/cm, 2 x 10 m/min
yellowness index YI: ASTM E 313 (formerly ASTM D 1925)

Table 4.8 shows the dependence of yellowing of a clear coat 15 minutes and one hour after curing on the photoinitiator and UV absorber. In every case, the lowest yellowness index YI is obtained with HPT-1 versus those samples without UV absorber or with BTZ-8. PI-II/PI-III in a 95:5 ratio causes less yellowing than PI-II/PI-III in a 3:1 ratio and only slightly more than with PI II alone.

These tests indicate that hydroxyphenyl-s-triazines are the preferred class of UV absorber for UV-curable clear coats. As for the photoinitiators, better results are obtained with blends of α-hydroxyketones (PI-I or PI-II) and bis-acyl phosphine oxides (PI-III) than for α-hydroxyketones alone. Of course, another important factor is the reactivity of the acrylate resins themselves [147, 148].

The medium-pressure mercury (Hg) lamps used for the experiments discussed produce a wide spectrum of radiation, including significant emissions in the UV region. Their broad emission spectrum permits the development of photoinitiators specific for targeted applications. One drawback of such mercury lamps is their strong emission in the IR. This produces a significant amount of heat that is highly problematic for plastic substrates, for example. Cooling through ventilation is possible but this requires energy. The last few years has witnessed the development of light emitting diodes (LED) emitting in the UV without the generation of heat. LED lamps are also more economical compared to the traditional UV lamps and should therefore pave the way for broader adoption of UV-curable resins in the future [149].

4.2.3 Specific requirements on HALS in coatings

Like UV absorbers, HALS (sterically hindered amines) used in coatings must meet certain requirements [23]. These are:

- maximum effectiveness
- very good solubility
- easy incorporation into standard coating formulations: in solvent-based coatings, HALS should be soluble in all commonly used solvents; in water-borne coatings HALS should preferably be a liquid and not require the use of a co-solvent; in radiation-curing coatings, the HALS should ideally be a liquid; in powder coatings, HALS should be a solid with a melting point in the same range as the extrusion temperature
- good compatibility/solubility in the polymers used, no exudation from the polymer network
- thermal stability
- low volatility between 80 °C and 150 °C
- no unwelcome interactions with other formulation ingredients such as catalysts
- extraction and migration resistance.

As with UV absorbers there is no general purpose HALS which meets all these requirements and each of the commercially available products on the market represents a compromise for any given type of application.

Some of the more important properties of these substances will be examined in the following Chapters, in somewhat less detail as the background for the different requirements has already been discussed in connection with UV absorbers.

4.2.3.1 Solubility and compatibility of HALS

Table 4.9 gives an overview of the solubility of different HALS. The liquid products are highly soluble in common solvents, ≥ 50 % and, thus, can be easily incorporated in all formulations containing organic solvents.

Liquid HALS can also be incorporated into liquid, solvent-free systems such as radiation-curing systems as well as some water-borne formulations. In certain water-borne systems it is sometimes preferred to incorporate the HALS in the organic phase before the water is added. Unless specific water compatible products or product forms (see Chapter 4.2.2.1.1) are being used it is advisable to carry out preliminary tests to find out the best way of incorporating the additive.

Some of the solid compounds in Table 4.9 are sparingly soluble, for example HALS-4 and HALS-9, and must first be dissolved in a suitable solvent before being added to the formulation. Due to their poor solubility, these compounds are not suitable for water-borne

Table 4.9: Solubility of HALS in various solvents

HALS	**Solubility (% w/w) @ 20 °C**			
	Methyl ethyl ketone	**N-butanol**	**Butyl acetate**	**Xylene**
HALS-1	24	40	n.a	24
HALS-2	> 50	> 50	> 50	> 50
HALS-3	> 50	> 50	> 50	> 50
HALS-4	9	2.5	10	10
HALS-5	20	25	na	20
HALS-6	> 50	> 50	> 50	> 50
HALS-7	> 50	> 50	> 50	> 50
HALS-8	> 50	> 50	> 50	> 50
HALS-9	16	0.4	n.a	10
HALS-10	> 50	n.a	> 50	> 50
HALS-11	> 50	> 50	> 50	> 50
HALS-12	> 50	> 50	27	7

or liquid radiation-curable coatings. In general, solid HALS are only suitable for powder coatings.

As with UV absorbers, HALS used in powder coatings should have a melting point in the range of the extrusion temperature of the coating in order to ensure a homogeneous distribution in the formulation. Any incompatibility of HALS with the resins will usually become apparent after curing.

HALS of very low polarity such as HALS-3 may exude from (very) polar solvent based formulations (for example polyester urethanes, epoxy/carboxy coatings) resulting in a drop of the initial gloss. In order to avoid this issue, other HALS such as HALS-10 or HALS-11 can be employed. Exudation of HALS with formation of a solid deposit on the coating surface is an exception and can be readily checked by applying the coating to a black surface followed by heat ageing, for instance at 60 to 80 °C for 24 h. Any incompatibility would be revealed through the formation of a deposit on the coating surface.

4.2.3.1.1 Water-borne systems

As already discussed in Chapter 4.2.2.1.1 the use of light stabilizers in water-borne coatings requires that they are easily dispersible in these formulations. The incorporation

Table 4.10: Examples for water compatible HALS

HALS	Technology	Chemical composition
HALS-13 [150] (basic)	solid state dispersion	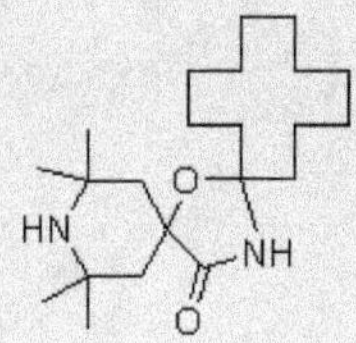
HALS-14 [150]	solid state dispersion	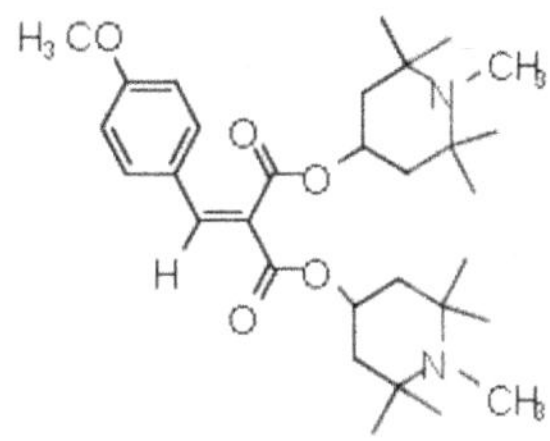
HALS-15 [150]	solid state dispersion	oligomeric HALS
HALS-16	nano dispersion	HALS-3

Table 4.11: Examples for water compatible UV absorber/HALS blends

LS-Blend	Technology	Chemical composition
LS-Blend-1	solid state dispersion	BTZ-5/HALS-13 (2/1)
LS-Blend-2	hydrophilic	BTZ-7/HALS-2 (2/1)
LS-Blend-3	nano dispersion	HPT-6/HALS-3 (2/1)

methods are essentially the same as described for UV absorbers. Nevertheless, specific water compatible HALS product forms mainly comprising solid state dispersions, emulsions and nano dispersions (NEAT) have been developed in recent years. All these technologies and the products have characteristic properties already described in greater detail in Chapter 4.2.2.1.1.

Since many applications, particularly clear coat applications require the combined use of UV absorbers and HALS for optimum protection, specific UV absorber/HALS blends have been developed. The typical UV absorber/HALS-ratio in these blends is 2:1 (g/g), representative for the vast majority of clear coat applications. Examples for water compatible HALS and UV absorber/HALS blends are shown in Tables 4.10 and 4.11 [128, 129, 150].

Table 4.12: Volatility of HALS products

HALS	% loss after .... minutes			
	15	**30**	**45**	**60**
HALS-2	0.4	0.6	0.7	0.9
HALS-3	0.7	1.0	1.2	1.3
HALS-4	0.1	0.1	0.1	0.1
HALS-7	0.4	0.5	0.5	0.5
HALS-9	0.2	0.2	0.2	0.3
HALS-10	0.2	0.3	0.3	0.4
HALS-11	0.2	0.4	0.5	0.6

Method: thermogravimetry (neat substance), isothermal at 120 °C

4.2.3.2 Volatility of HALS

HALS, like UV absorbers, may be exposed to high temperatures during the curing process. As for UV absorbers the loss of HALS at high temperatures can be prevented by an increase in molecular weight: loss of HALS-9 is for example almost completely suppressed by its high molecular weight. However, an increase in molecular weight may lead to compatibility problems, depending on the resin used.

It is not an easy matter to determine the volatility of HALS in coatings since the loss of these substances cannot be easily monitored spectroscopically as for UV absorbers. One option is the solvent extraction followed by HPLC analysis. This method, however, only works prior to the formation of the active HALS form, the nitroxyl radical (see also Chapter 4.2.3.3).

A quicker method of determining the volatility of a given substance is thermogravimetry. Table 4.12 shows the weight loss of various HALS products after isothermal heating to 120 °C. All these suffer very little weight loss (< 1.5 %) even after one hour.

Due to the molecular weight of the compounds and the steric conditions prevailing in the three-dimensional network of the polymer, it can be assumed that the actual losses under normal conditions are even less. Migration (see also Chapter 4.2.3.1.) is more or less independent of volatility and can take place in two directions: toward the surface of the coating system (exudation) or through the lower layers of the complete film build up.

Since migration can be accompanied by reduced or even loss of the stabilization effect, the use of reactable HALS may be necessary or even mandatory in certain applications.

4.2.3.3 Reactable HALS

The important differences between HALS-3 that can react with certain binder components and the non-reactable HALS-10 (primary OH in side chain, see Table 3.3 in Chapter 3.3.2) are revealed in Figure 4.16 [131, 132]. The experimental set-up is the same as for testing UV absorbers: 1 % based on binder solids of each of the compounds was incorporated in both the clear coat (2P-PU) and the basecoat. The clear coat/basecoat system was applied to two different substrates, steel or a thermoplastic polyolefin (TPO) widely used in the automotive industry. The HALS content was determined by extraction of the microtomed coating Chapters. The difficulties in the analysis of HALS required the development of a specific analytical method which involves an in-situ conversion of all HALS moieties into the corresponding nitroxides that can be quantified by ESR analysis. The concentration of HALS based on binder solids can then be derived the nitroxide concentration [151, 152]. In the present case, analysis also included the TPO substrate.

The results depicted in Figure 4.16 show a more or less even concentration of the reactable HALS-10 throughout the clear and basecoat after 3000 h exposure, which indicates that no migration occurred. In case of HALS-3 a significant loss in the upper layer is apparent. It can be even detected in the TPO substrate down to about 300 μm.

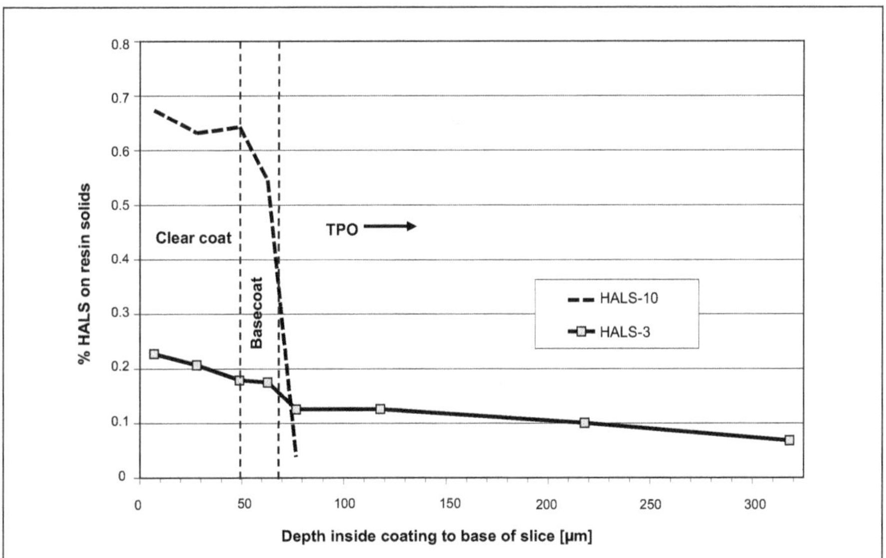

Figure 4.16: HALS concentration vs depth profile in model 2P-PU clear coat
basecoat: Prairie Tan Metallic, substrate: TPO, exposure: 3000 h exposure A

The conclusion is, not surprisingly, the same as for UV absorbers: coatings over plastics should be stabilized with reactable additives to ensure good protection of the entire system.

4.2.3.4 Effect of HALS on coating colour

Since HALS do not absorb light above 250 nm (unless they have been modified), they appear colourless. It is nevertheless advisable to carry out preliminary tests, as in the case of UV absorbers, to ascertain their behaviour in a formulated system. Figure 4.17 shows the result of one of such tests. A 2P-PU clear coat (hydroxyl-functional acrylate/isocyanate) containing 1.5 % HALS was applied to a white basecoat in two different thicknesses and baked for 30 minutes at 90 °C. The yellowness index b*, determined immediately after baking as well as after 1 hour aging at 90 °C, show that none of the HALS had any noticeable effect on the colour of the coating. In addition, HALS, unlike UV absorbers (see Table 4.5), do not cause increased yellowing with increasing film thickness: when the coating thickness is doubled from 40 to 80 µm, the increase in b* was not greater than in the coatings without HALS. Negative b* values indicate a shift on the yellow/blue axis into the blue.

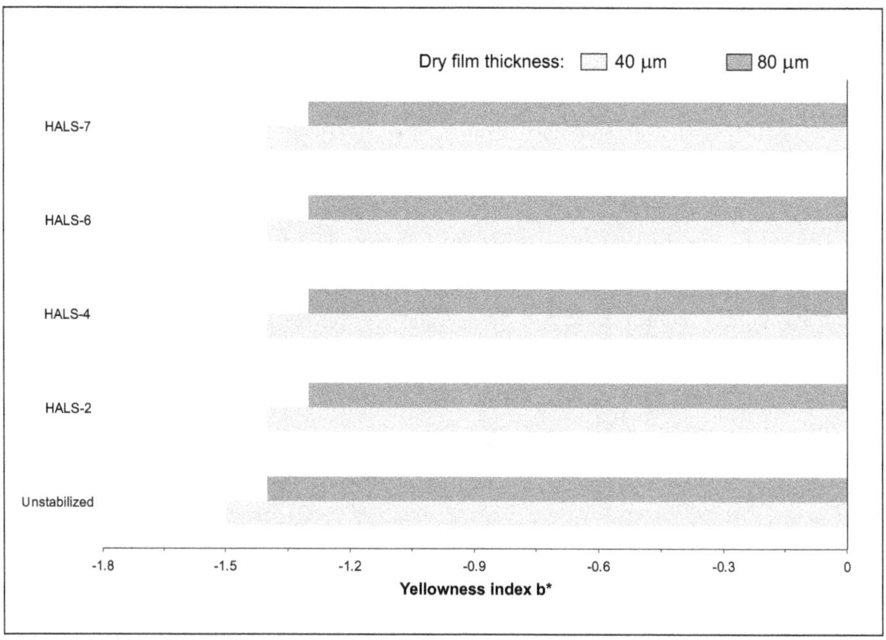

Figure 4.17: Effect of HALS on the yellowing of a 2P-PU clear coat
basecoat: white, bake: 90 °C, 30'

Table 4.13: pk_b values for various HALS products[1]

general structure:

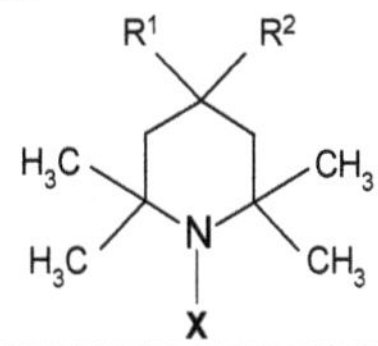

HALS	X	pk_b
HALS-1	H	~ 5.0
HALS-2	CH_3	~ 5.0
HALS-3	OC_8H_{17}	~ 9.6
HALS-4	CH_3	~ 5.3
HALS-5	$C(O)CH_3$	> 12.0
HALS-6	H	~ 5.0
HALS-7	CH_3	~ 5.3
HALS-8	$C(O)CH_3$	~ 11.5
HALS-9	CH_2CH_2OCOR	~ 7.5
HALS-10	O-cyclohexane	~ 9.4[1]
HALS-11	$(CH_2)_2OC(O)CH_2CH(CH_3)CH_2$-tert. butyl	~ 8.0

1 pk_b of piperidine N

4.2.3.5 Unwelcome side reactions

The stabilizing effect of HALS emanates from the tetramethylpiperidine group. Thus, the key properties of HALS can be adjusted by changing the substituents of the nitrogen atom, fine-tuning them for specific applications. Here, two factors are especially important:

a. the basicity of the HALS
b. the activation energy (speed) necessary to convert the HALS into the active substance (nitroxyl radical).

Factor b) has already been discussed in Chapter 3.3.2.1 and is not relevant to possible unwelcome side reactions.

The basicity of HALS determines to a great extent the applications in which is can be used. Table 4.13 shows the pk_b values, a measure of basicity, of various HALS. In general, HALS carrying a hydrogen or methyl group on the piperidine nitrogen are distinctly basic and have a pk_b value of approximately 5. HALS with $COCH_3$ or O-alkyl, for example OC_8H_{17} substituents are not basic.

Table 4.14: Effect of HALS on the pendulum hardness of an acid-catalyzed HS-TSA clear coat

general structure: 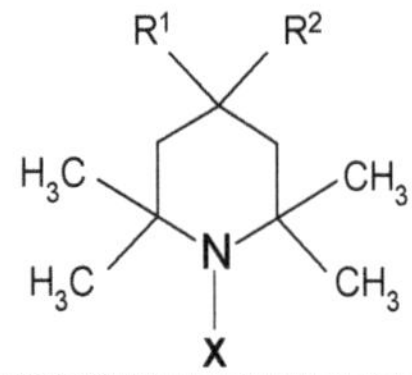

HALS (1 %)	X	Pendulum hardness [s] after bake
Without	-	158
HALS-1	H	3
HALS-2	CH_3	14
HALS-3	OC_8H_{17}	158
HALS-5	$C(O)CH_3$	162
HALS-8	$C(O)CH_3$	158
HALS-9	CH_2CH_2OCOR	118

Basecoat: white, bake 120 °C, 30'

Table 4.15: Comparison of the effect of various HALS products on the pendulum hardness of an acid-catalyzed HS-TSA clear coat

general structure: 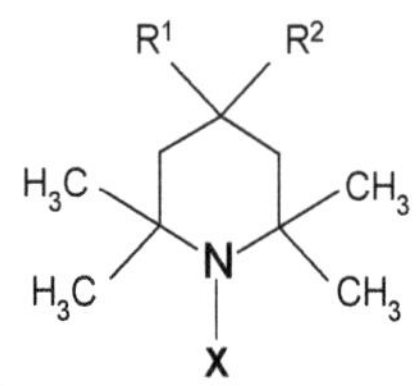

HALS	X	Pendulum hardness [s] after bake
Without	-	122
1 % HALS-2	CH_3	sticky – cannot be measured
1.4 % HALS-3	OC_8H_{17}	130
1.5 % HALS-5	$C(O)CH_3$	124

Substrate: steel, bake 130 °C, 30'

Basic properties can have adverse effects in the following formulations:

a. **base catalyzed coatings**: the crosslinking reaction can be further accelerated by a basic HALS leading to shorter pot lives. This applies particularly to the hydrogen substituted HALS, HALS-1 and HALS-6.

b. **water-borne systems** which have been neutralized with amines: the basic HALS may cause a drastic change in the pH value so that the liquid formulation may be destabilized or even coagulate
c. **acid-catalyzed coatings** such as HS-TSA and formulations with a high acid number: the basic HALS will interact with the acid and/or the resin. If the HALS interacts with the catalyst , curing inhibition or an undercured coatings will result.
d. **oxidative curing systems** (for example alkyd systems as being used for example in industrial or decorative applications): a basic HALS may interact with the metal driers leading to cure retardation or prolonged drying times.

Non-basic HALS such as HALS-3, HALS-5 and HALS-8 generally do not show these limitations. These HALS were initially developed for acid-catalyzed coating systems [153–157]. It is always advisable to carry out preliminary tests to identify the most appropriate HALS for an application.

Tables 4.14 and 4.15 show the effect of different HALS on the pendulum hardness of two acid catalyzed HS-TSA clear coats. In Table 4.14, different HALS are compared at identical concentrations (1 %, calculated on clear coat solids) in a coating catalyzed with 0.5 % p-toluene sulphonic acid. The pendulum hardness according to König (DIN 53 157) was used as test criterion. The results show quite clearly that only coatings containing HALS-3, HALS-5 and HALS-8 exhibit a pendulum hardness comparable to the clear coat without HALS. The pendulum hardness values obtained with the other HALS are low, indicating undercuring or even complete cure inhibition. Clearly, the basic HALS neutralize the acid needed to catalyse the crosslinking reaction to yield ammonium salts. Such coatings are not only undercured but also moisture sensitive because of the salt formed [153].

Table 4.15 shows an additional comparison of basic and non-basic HALS in an acid-catalyzed HS-TSA clear coat. The HALS were evaluated at identical concentrations of the active piperidine nitrogen. As expected, in contrast to HALS-2, HALS-3 and HALS-5 had no influence on the acid-catalyzed crosslinking reaction.

4.2.4 Weathering results for two-coat systems

This Chapter discusses the effects of UV absorbers and HALS on different coatings and varnishes. First, however, let us briefly look at test methods and assessment criteria.

4.2.4.1 Weathering tests

Here one must distinguish between natural weathering *(outdoor weathering)* and artificial weathering *(accelerated weathering)*. The purpose of accelerated weathering is to obtain a quick indication of weatherability as outdoor tests may require up to several years for high quality coatings. Clearly, substantial agreement between the accelerated and outdoor

Table 4.16: Outdoor exposure tests

Location/country	Exposure conditions	Abbrevation
Miami area/USA	5° facing south, black box, unheated	Florida (5° S, bb, unheated)
Miami area/USA	45° facing south, black box, unheated	Florida (45° S, bb, unheated)
Allunga/Australia	5° facing north, black box, unheated	Allunga (5° N, bb, unheated)
Basle/Switzerland	45° facing south	Basle (45° S)
Sydney/Australia	45° facing north	Sydney (45° N)

weathering tests would be ideal. Unfortunately, an unequivocal relationship between the two types of test cannot be established rendering the lengthy outdoor tests indispensable. Thus, the development and introduction of new light stabilizers is a long-term endeavour.

In **outdoor weathering tests**, the closer the chosen weathering conditions are to those likely to be encountered in use, the more useful the results obtained. Outdoor weathering tests can be conducted in various regions of the world, whereby the full range of potential climatic conditions are accessible. Typical weathering stations include:

- the Alps (high UV radiation)
- the North Sea (salty air)
- the Ruhr region (industrial atmosphere)
- Jacksonville, USA (industrial atmosphere)
- Florida, USA (subtropical climate)
- Arizona, USA (desert climate)
- Allunga, Australia (increased UV radiation, small climatic variations, tropical)

In order to close the gap between accelerated and natural weathering in terms of the duration of the exposure period needed, *Atlas Material Testing Technology LLC* has additionally pioneered accelerated outdoor weathering tests known as "Emma" (Equatorial Mount with Mirrors for Acceleration) or "Emmaqua" (Equatorial Mount with Mirrors for Acceleration, with water (aqua)). The sun tracking devices used thereby concentrate natural sunlight via 10 flat mirrors onto the specimens resulting in approximately 5-fold increase in UV radiation. Testing of automotive coatings is typically performed according to ASTM G 90 (cycle 3), which includes night time wetting, e.g. three minutes' water spray per hour in between 7 pm and 5 am. Obviously, the weathering conditions chosen must be in line with the intended application. For the car industry, the results of weathering tests carried out in Florida are the most relevant. Table 4.16 gives an overview of outdoor weathering tests carried out on coatings, which will be discussed in more detail later.

Because of the extreme climatic conditions in Florida and Allunga, weathering tests carried out in those places take far less time than those done in Central Europe. Test panels are arranged in racks facing 5° South (Florida) and 5° north (Allunga), so that they are exposed to maximum solar radiation. Furthermore, these weathering tests are carried out under so-called black box conditions: the test panels are placed on metal boxes which are painted black on the outside. These boxes are closed and are heated only by the sun, not by any other external heat source. This more or less simulates the weather conditions to which a horizontal car body component is exposed in summer. Further information on outdoor weathering tests is available in the literature.

In choosing a suitable accelerated weathering test, the type of coating, the method of application (one-coat or two-coat) and the expected/required damage to the coating must be taken into account. Table 4.17 summarizes the most common accelerated weathering tests used by the car industry. Figure 4.18 shows the emission spectra of the lamps employed, whilst Table 4.18 lists the most important criteria for the assessment of coating properties.

UVB fluorescent lamps with a maximum intensity at approximately 313 nm (UVB-313 lamps) have been used in weathering tests since the early seventies. These lamps emit

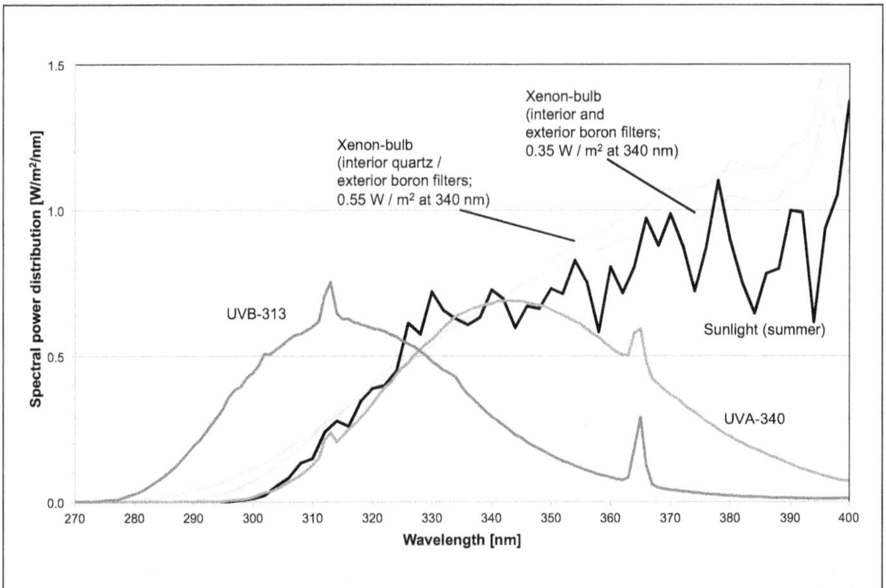

Figure 4.18: Comparision of sunlight to the UV emission spectra of lamps most widely used in the accelerated weathering tests Source: Q-Lab Corporation, 800 Canterbury Rd, Westlake, OH 44145, USA

Table 4.17: Overview of accelerated weathering tests

Instrument (manufacturer)	Type of lamp	Cycle/purpose	Abbrevation exposure/ weathering conditions
Xenon-Weather-o-meter (Atlas[1], Q-Lab Corporation)	Xenon Arc	SAE-J 2527 (formerly SAE-J 1960)[a]/ Automotive exterior	A
Xenon-Weather-o-meter (Atlas[1])	Xenon Arc	DIN EN ISO 16474-2 (replaces DIN EN ISO 11341:2004-12)[b]/ Exterior paints & varnishes (Automotive, Industrial, Wood, plastics)	B
QUV (Q-Lab Corporation), UVCON (Atlas[1])	Fluorescent lamps UVB-313	ASTM G 154 Cycle 2[c]/ Standard practice for non-metallic materials	C
	Fluorescent lamps UVB-313	ASTM G 154 Cycle 1[d]/ Standard practice for non-metallic materials	D
	Fluorescent lamps UVA-340		E
Condensing Humidity Cabinet (e.g. Q-Lab Corporation)	-	24 to 96 h, 38 °C, 100 % r.h./humidity (water) resistance	F
High temperature light exposure (Atlas[1])	Xenon Arc	PV 1303 (VW, Audi)[e]/ Automotive interior (coatings, plastics, elastomers, textiles)	
Daylight exposure in warm & humid climates, e.g. Florida (Atlas[1], Q-Lab Corporation)	Xenon Arc	PV 3390 (VW)[f]/ Automotive exterior (plastics, elastomers)	
Daylight exposure in hot & dry conditions, e.g. Kalahari, Arizona, South Africa (Atlas[1], Q-Lab Corporation)	Xenon Arc	PV 3929 (VW)[g]/ Automotive exterior (coatings, plastics, elastomers, convertible roof covers)	
Super UV Tester (Iwasaki Electric Co. Ltd.)	Metal halide	Irradiance 100 x greater than natural UV, 10 x faster than e.g. Xenon exposure (mainly used in Far East)/Automotive exterior	
QUV (Q-Lab Corporation)	Fluorescent lamps UVA-340	DIN EN 927-6[h])/ Exterior wood coatings	G

Table 4.17: Overview of accelerated weathering tests (continuation):

1 Atlas Material Testing Technology

a 0.55 W/m^2 at 340 nm; quartz inner filter, type S borosilicate outer filter. 120 min exposure to light (40 min dry, 20 min rain on front side of sample; 60 min dry); 60 min dark (rain front or back of sample). Black panel temperature 68–72 °C (light), 35–43 °C (dark).

b 0.35 W/m^2 at 340 nm; type S borosilicate inner and outer filter. 102 min exposure to light (dry); 18 min exposure (rain). Black panel temperature: 57–63 °C (dry), 37–43 °C (rain)

c 0.64 W/m^2nm at 313 nm; 8 h exposure to light at 60 °C, 4 h dark & condensation at 50 °C; alternative cycle: 4 h exposure to light at 60 °C, 4 h dark & condensation at 50 °C

d 0.77 W/m^2nm at 340 nm; 8 h exposure to light at 60 °C, 4h dark & condensation at 50 °C

e 60 W/m^2 at 300–400 nm (dry); black panel temperature 100 ± 3 °C, specimen room temperature 65 ± 3 °C, 20 ± 10 % r.h.

f 60 W/m^2 at 300–400 nm; 102 min. exposure to light (dry), 18 min exposure (dark, rain); black panel temperature 65 ± 2 °C; 60–80 % r.h during dry cycle

g 75 W/m^2 at 300–400nm (dry); black panel temperature 90 ± 2 °C, specimen room temperature 50 ± 2 °C, 20 ± 10 % r.h.

h 0.87 W/m^2 nm at 340 nm; stage 1: condensation (dark) at 45 ± 3 °C; stage 2: 144 h consisting of 48 cycles each lasting for 3 h and comprising 2.5 h light at 60 ± 3 °C and 0.5 h rain (dark); stage 3: spray (6–7 l/min., dark)

at wavelengths not present in natural sunlight, thereby exposing the coating to high energy and greatly accelerating binder degradation. This type of weathering test was first thought to be suitable for assessing the cracking tendency of clear coats and that several thousand hours cracking resistance under UVB 313 conditions would translate into similar cracking resistance during natural exposure. However, experience has shown that this is not the case. The results regarding chalking or cracking of one-coat finishes or the discolouration of one and two-coat finishes do not agree with those obtained in outdoor weathering tests [158].

UVA fluorescent lamps with maximum intensity at about 340 nm (UVA-340 lamps) were introduced in 1987. Their emissions at wavelengths of between 300 and 360 nm matches well with natural sunlight. Furthermore, the weathering tests with these lamps require more time than with the UVB-313 lamps, indicating that they better simulate typical outdoor conditions that the latter. In 1992, modifications of the weathering instrumentation permitted a more uniform light intensity leading to shorter weathering times [159]. Indeed, it has been shown that UVA-340 lamps generate relatively reliable results with regard to gloss retention and cracking of two-coat finishes [160]. However, as these lamps do not emit in the visible, they are only of limited value in the assessment of colour changes arising from the pigments in coloured coatings.

Xenon lamps equipped with special filters best simulate natural light and allow the closest correlation of accelerated weathering to outdoor weathering. Use of these lamps and the typical variation of conditions, light and dark phases and as well as dry and wet phases

Table 4.18: Assessment criteria for coating films

Criterion	Standard
Gloss (20°, 60°, 85°); unpolished	DIN 67530
DOI (Distinctness of reflected Image)	ASTM E 430
Orange peel/wave-scan DOI	Customer/OEM specific
Cracking	TNO-scale[1]
$L^*a^*b^*$ colour space (CIELAB)	DIN EN ISO 11664-4 (formerly DIN 6174)
Yellowness index b^*	DIN EN ISO 11664-4 (formerly DIN 6174)
Yellowness index YI	ASTM E 313 (ASTM D 1925, withdrawn but still widely used)
Pendulum hardness (König)	DIN 53157, ASTM D 4366-14
Cross-hatch adhesion	DIN EN ISO 2409:2013-06
Stone chip resistance	DIN EN ISO 20567-3

1 Toegepast Naturwettenschaffplijk Onderzoek (TNO Centre for Coatings Research, Delft/NL)

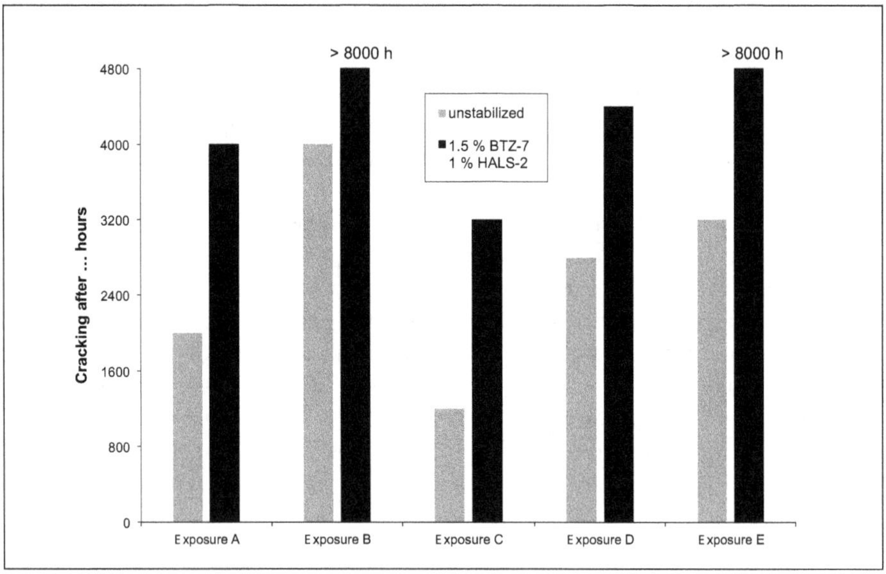

Figure 4.19: Effect of exposure on cracking
clear coat: water-borne, acrylate/melamine, basecoat: water-borne, silvermetallic, bake: 70 °C, 10' + 130 °C, 20'

and varying temperatures, allow a good assessment of the weathering of one and two-coat finishes with regard to gloss retention, cracking, changes of colour and adhesion. Similar to the UVA-340 lamps, xenon lamps produce only small amounts of the intensive short-wave radiation that causes polymer degradation so the weathering tests must be carried out over a longer timeframe.

Clearly, the selection of the lamp is critical to obtain trustworthy accelerated weathering test results but the lamp is not the only concern: weathering is also influenced by moisture and temperature changes. The most important parameters are summarized in Table 4.18.

To illustrate this fact, Figures 4.19 and 4.20 show the weathering of a two-coat (Figure 4.19) and a one-coat finish (Figure 4.20) using different weathering instruments.

Figure 4.19 shows the time needed to produce cracks in an unstabilized, water-borne clear coat compared to an identical clear coat stabilized with a blend of BTZ-8 and HALS-2. Under weathering conditions C (UVB-313 lamps), cracking of the coating was observed much earlier than under weathering conditions B (xenon lamp) and E (UVA-340 lamp). The UVB-313 lamp (C) has a greater intensity at shorter wavelengths than the xenon lamp (B, cf. Figure 4.18) where the polymer is more sensitive. The unstabilized clear coat cracks earlier in all cases.

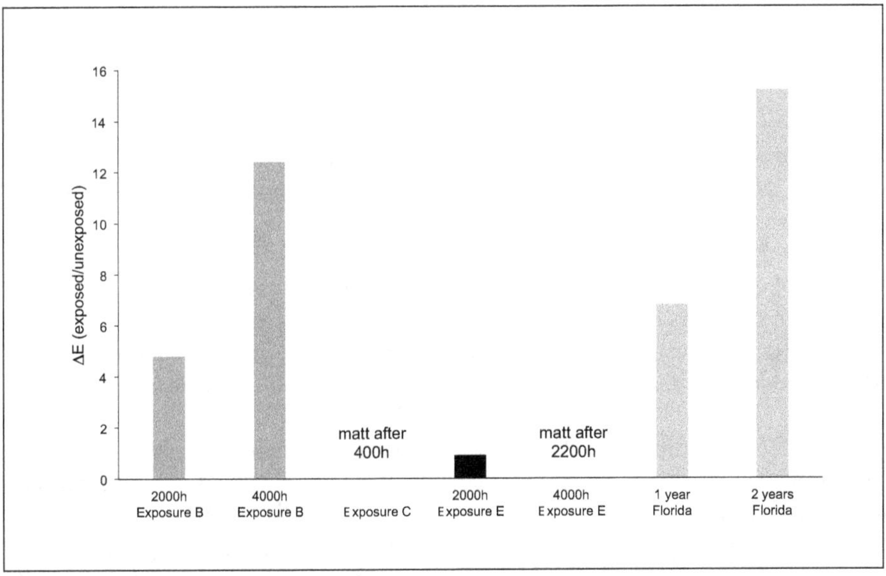

Figure 4.20: Comparison of different accelerated weathering conditions with outdoor exposure of an unstabilized 1-coat system

binder: alkyd/melamine, pigment: Pigment Red 178/TiO_2 (50/50), bake: 130 °C, 30'

Weathering conditions A and B make use of a xenon lamp but in A, a different filter system is used that allows the passage of higher light intensity. In addition, higher temperatures and more moisture (change from light to dark and dry to wet) are present. Under conditions A, cracks take longer to form than under conditions C, but distinctly less time than under conditions B and E. In Figure 4.20 weathering conditions B, C and E are compared with outdoor weathering in Florida (5° South, bb, unheated).

The tests were carried out using a red pigmented alkyd/melamine coating applied as a one-coat finish and the weathering was assessed based on colour change, ΔE. Quite obviously, weathering conditions C are totally unsuitable for this particular system. The intense emission at 313 nm causes very rapid degradation of the binder, resulting in a matt and chalky surface. Weathering conditions E are not much better. The closest agreement with outdoor weathering results is provided by weathering conditions B (xenon lamp).

Several analytical methods have been developed for quantifying polymer degradation in binder systems. These methods involve analysis of the chemical changes taking place in the polymer matrix [85, 143, 161–166]. Some of these chemical changes occur very early during weathering and therefore allow a prediction of long-term weathering properties. The drawback of these methods is the complexity and, in certain cases the cost.

4.2.4.2 Results for solvent-borne clear coats

Let us first re-emphasize the need to stabilize clear coats with a blend of UV absorber **and** HALS. In this study, as examples three clear coat systems were applied onto a basecoat.

Figure 4.21 shows the effect of HALS on gloss retention in a high-solids 2P-PU clear coat. Coatings stabilized with only a UV absorber showed cracking after 30 to 33 months. The addition of 1 % HALS-2 prevented cracking of any of the coatings over more than 9 years. The gloss decreased to 26 and 48 units from an initial value of 93. No decrease in adhesion was observed for the coatings containing BTZ-8 and HPT-1.

Figure 4.22 and Table 4.19 show the additional influence of UV absorber on gloss retention, colour retention and adhesion. As Figure 4.22 shows, the addition of various HALS to an acrylic/melamine clear coat results in a decrease in gloss from 90 to 20 or 45 units after 4000 hours of weathering under conditions A. If 1.5 % of the UV absorber BTZ-8 is added, the decrease in gloss is minimal, values of 80 units are measured after 4000 h.

Table 4.19 shows the effect of UV absorbers on colour retention and adhesion. UV absorbers prevent or delay degradation of both the binder material and pigment. Degradation of the binder in the basecoat can adversely affect the adhesion of the clear coat to the basecoat. Pigment degradation, on the other hand, leads to a colour change of the basecoat. If the clear coat is only stabilized with 1 % HALS, there will be a significant change of colour and reduced adhesion (Gt 4). In coatings containing 2.5 % UV absorber, the

Table 4.19: Effect of UV absorbers on coating properties of a 2-coat metallic coating

Stabilization	Color change ΔE (exposed/unexposed)	Adhesion (clear coat/basecoat)[1]
1 % HALS-2	27.8	Gt 4
1 % HALS-2 2.5 % BP-3	16.0	Gt 0
1 % HALS-2 2.5 % BTZ-8	8.2	Gt 0
1 % HALS-2 2.5 % HPT-1	7.9	Gt 0
1 % HALS-2 1.7 % BTZ-8 0.8 % HPT-1	6.4	Gt 0

clear coat: TSA basecoat: red metallic
bake: 130 °C, 30' exposure: 4000 h exposure B

1 According to DIN 53151 after additional exposure F; tape test

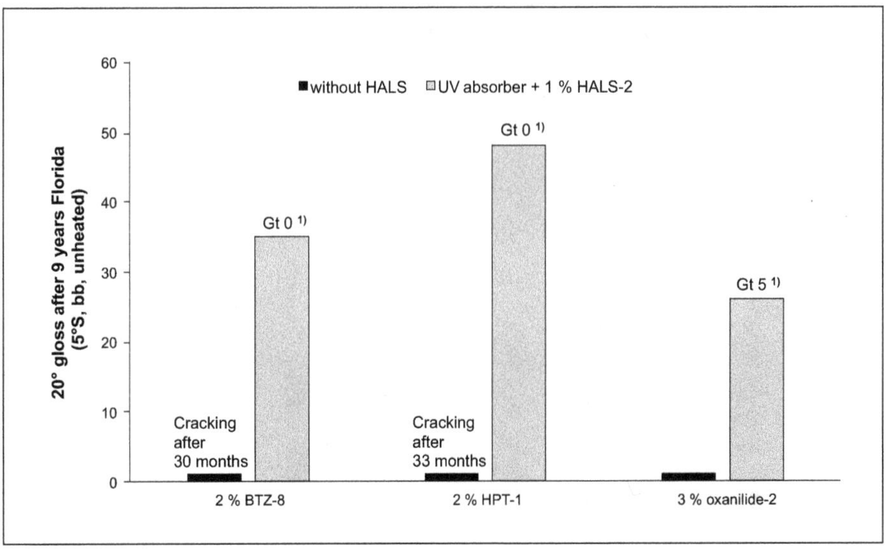

Figure 4.21: Effect of additional stabilization with HALS on gloss retention of a HS-2P-PU clear coat
basecoat: water-borne, silver metallic, bake: 90 °C, 30'

1 According to DIN 53151 after additional exposure F; tape test

adhesion was unaffected by 4000 hours' weathering under conditions B. Depending on the type of the UV absorber used, the colour retention was significantly improved as well.

The colour retention and improvement in adhesion are directly attributable to the screening or filtering effect of the UV absorbers (see Table 4.19).

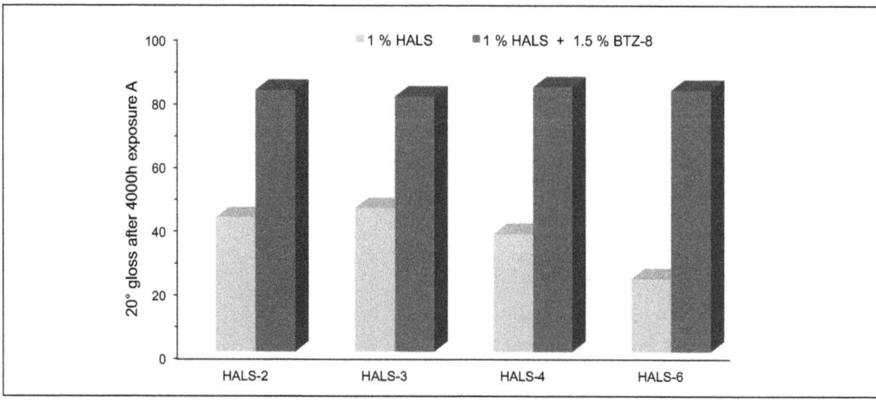

Figure 4.22: Effect of UV absorber on gloss retention of a TSA clear coat
basecoat: silver metallic, bake: 130 °C, 30'

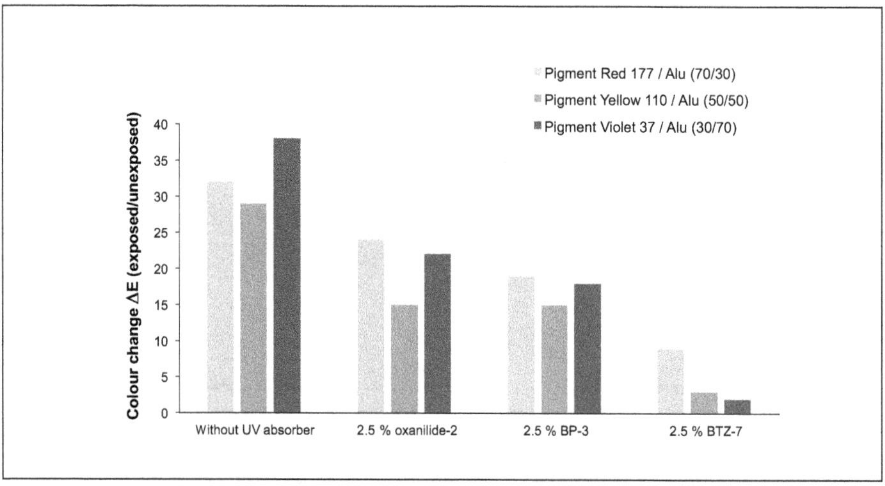

Figure 4.23: Filter efficiency of different UV absorber types: basecoat discolouration after 54 months' outdoor exposure
clear coat: TSA, basecoat: metallic, bake: 130 °C, 30', exposure: Florida, 5° S, bb, unheated

Each Formulation contains additional 0.5 % HALS-2

As shown in Figures 3.4, 3.5, 3.13 and 3.14 of Chapter 3, the broadness of the absorption band of a UV absorber and, thus, its capacity to filter out harmful UV light, depends on its structure. The filtering efficiency of typical UV absorbers is illustrated in Figures 4.23 and 4.24

A comparison of different types of UV absorbers, an oxanilide, a benzophenone and a benzotriazole, in clear coats stabilized with 0.5 % HALS-2 (to prevent cracking) applied on three different pigmented basecoats is shown in Figure 4.23. After weathering in Florida for 54 months, it is apparent that HALS-2 alone does not prevent colour change of any of the basecoats. However, the use of UV absorber can lead, depending on the type of UV absorber used, to a significant improvement.

The best protection against colour change is provided by BTZ-7, which has the broadest spectral coverage and can, therefore, filter out most of the UV light. Both oxanilide-2 and BP-3 have an absorption band shifted to the short-wave range and allow greater penetration of UV light through the clear coat, and, thus cannot sufficiently protect the basecoat against discolouration.

Figure 4.24 shows a comparison of BTZ-3 to HPT-1 in the accelerated weathering of clear 2P-PU coating applied on a light blue metallic basecoat. Although HPT-1 has a little lower absorption at longer wavelengths than BTZ-3, less discolouration occurred.

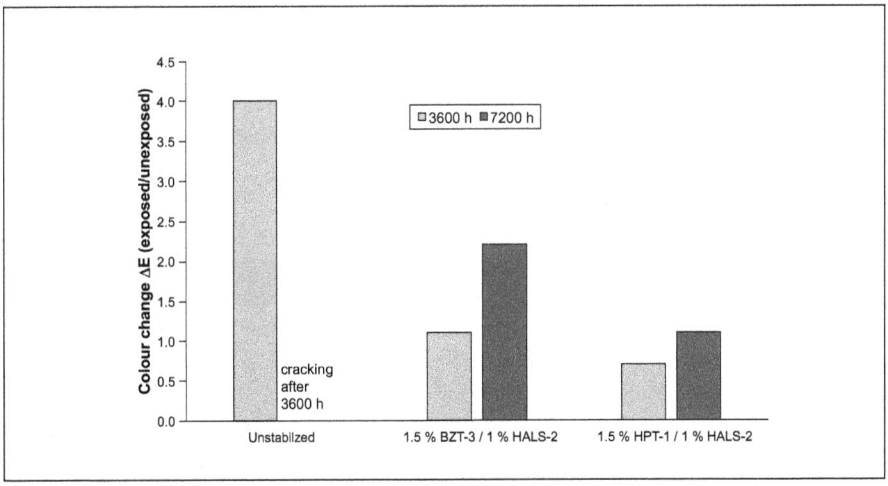

Figure 4.24: Colour retention of a 2-coat metallic coating after 3600 hours and 7200 hours exposure A
clear coat: 2P-PU, basecoat: light blue metallic, bake: 80 °C, 30′

The results from Table 4.19 and Figures 4.23 and 4.24 can be rationalized as follows:

- the spectral coverage (broadness of the absorption) bands is a measure for the filtering (screening) effect of UV absorbers;
- benzotriazoles and HPT are better filters of UV light than oxanilides and benzophenones;
- another critical wavelength of light which still has to be absorbed lies between the absorption of benzophenone and HPT;
- a blend of two UV absorbers can lead to further improvement of the stabilizing effect, depending on the pigment. In this case, the absorption of the UV absorbers must be complementary (see Table 4.19: 1.7 % BTZ-8 + 0.8 % HPT-1).

Although the filter effect of a UV absorber is of great importance, its stability in the coating cannot be disregarded. This will be discussed in more detail in Chapter 4.2.6.

HALS are an important component of the stabilization package. As aforementioned in Chapter 3.3.2.1, the piperidine N substituent of HALS has a great influence on the rate of its conversion into the active nitroxyl radical >N-O· and, therefore, a decisive influence on its stabilizing effect. Laboratory tests have shown that >N-CH_3 and >N-O-C_8H_{17} are converted more quickly into >NO· than >$NCOCH_3$ [157]. The question remained whether faster conversion actually does result in improved stabilization under normal conditions. The proof was obtained from trials portrayed in Figures 4.25 to 4.27 and Table 4.20. Figure 4.25 compares the cracking resistance of a clear, acid catalyzed, high-solids acrylic/melamine coating cured for 30 minutes at 120 °C containing differently substituted HALS-3 and HALS-5.

A coating containing 1 % HALS-3 (>N-O-$_8H_{17}$) in combination with 2 % BTZ-7 shows far greater resistance to cracking under accelerated weathering conditions than analogous coatings containing combinations of 1 % or 2 % HALS-5 (>N-$COCH_3$) with 2 % BTZ-7. Figure 4.26 shows the 5 years outdoor weathering results for another high-solids acrylic/melamine clear coat that was cured under car refinishing conditions (30 min at 90 °C).

The gloss retention of the unstabilized sample was poor and crack formation occurred after two years. Of the HALS-stabilized coatings, that containing HALS-3 (>N-O-C_8H_{17}) retained higher gloss and was less susceptible to cracking than that with HALS-5 (>N-$COCH_3$). Even at twice the concentration, the latter developed cracks after 4 years and exhibited only slightly better gloss than the coating with only 1 % HALS-5.

Figure 4.27 is a comparison of HALS-5, HALS-8 (both >N-$COCH_3$ derivatives) and the N-O-C_8H_{17}- substituted HALS-3 in combination with different UV absorbers.

After 5 years' outdoor exposure in Florida, the same acid-catalyzed high-solids acrylic/melamine clear coat containing 2 % HPT-1 and 1 % HALS-3 exhibits better gloss retention than coatings stabilized with 2 % HPT-3 and 2 % HALS-8 or 2.5 % oxanilide-2 and 2 % HALS-5.

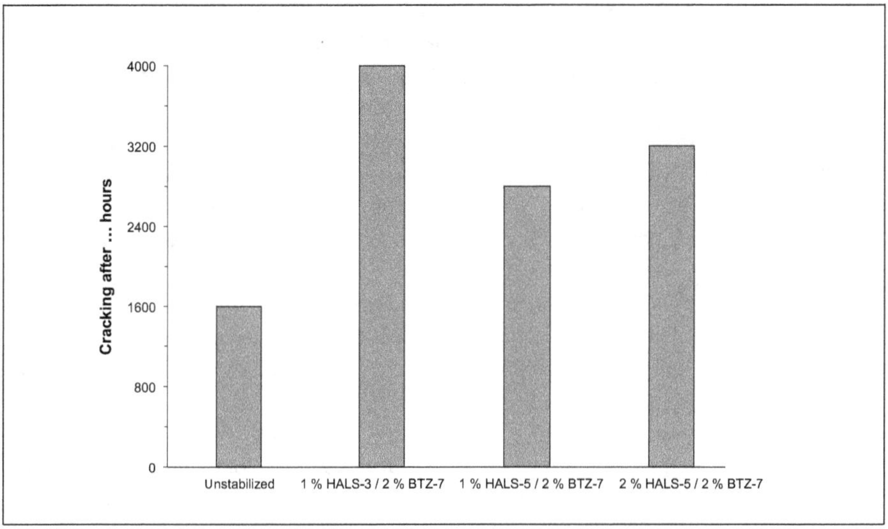

Figure 4.25: Effect of nitrogen substitution on the stabilizing effect of HALS (exposure D)
clear coat: HS-TSA, basecoat: silver metallic, bake: 120 °C, 30'

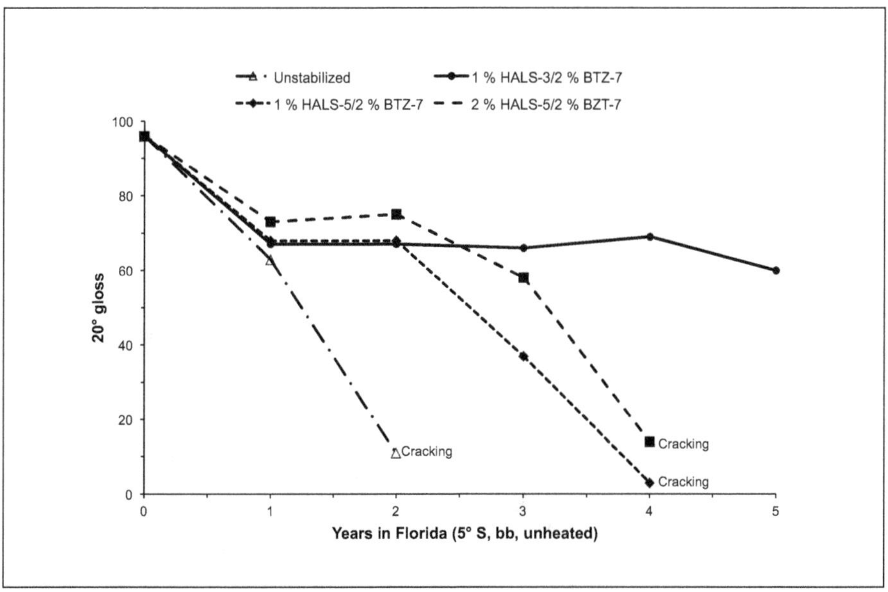

Figure 4.26: Effect of nitrogen substitution on the stabilizing effect of HALS
clear coat: HS-TSA, basecoat: silver metallic, bake: 90 °C, 30'

In model experiments [86], HALS with N-O-C_8H_{17} substituents are converted into the corresponding >NO· just as quickly as >N-CH_3 substituted HALS. This is also the case in prac tice. Table 4.20 compares coatings containing HALS-3 (>N-O-C_8H_{17}) and BTZ-7 to coatings containing HALS-2 and HALS-4 (both >N-CH_3) with the same UV absorber.

After 7 years' outdoor exposure in Florida, all coatings showed similar gloss retention. The adhesion after 5 years' outdoor exposure was perfect for all the stabilized formulations. Thus, the stabilizing effect of >N-O-R HALS is comparable to that of >N-H and >N-CH_3 substituted HALS. Note, however, that only the former can be added to systems that are base-sensitive.

The effect of some stabilizers on the glass transition temperature T_g of a high-solids acrylic/melamine clear coat is depicted in Table 4.21. A pronounced increase in T_g of a film indicates loss of flexibility, an increase in brittleness. High T_g films are more susceptible to damage, for example by stone chipping.

The glass transition temperature of the unstabilized sample increased by 24 °C after 500 hours' accelerated weathering whereas that of the sample containing only a UV absorber showed only a 9 °C increase. A further reduction to 5 °C is achieved by adding HALS. Thus, the degradation of the mechanical properties of coatings upon weathering can be prevented or at least greatly reduced by the use of light stabilizers.

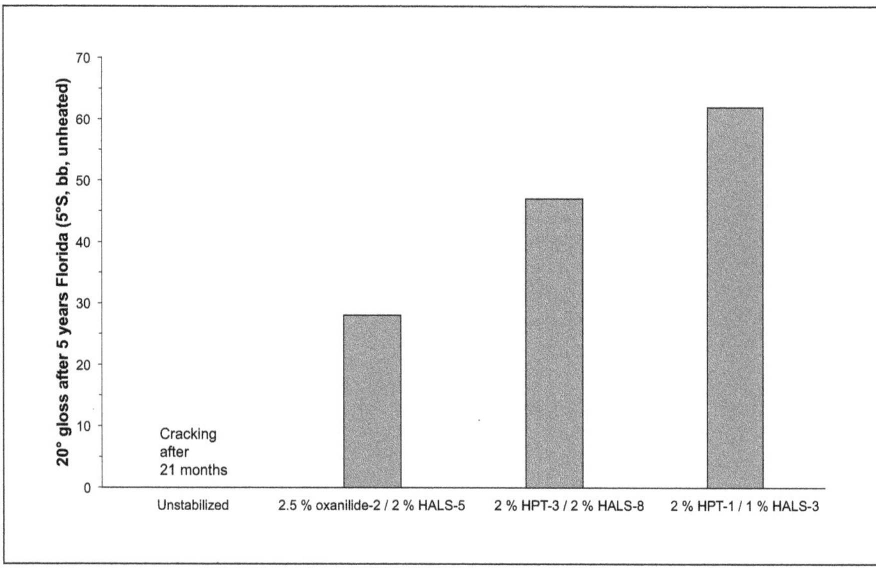

Figure 4.27: Comparison of various UV absorbers and HALS in a HS-TSA clear coat
basecoat: HS-TSA, silver metallic, bake: 120 °C, 30'

Table 4.20: Comparison of a >N-O-C_8H_{17} and >N-CH_3 substituted HALS

Stabilization	**20° gloss after ... years in Florida (5°S, bb, unheated)**								**Adhesion[1] after 5 years exposure**
	0	**1**	**2**	**3**	**4**	**5**	**6**	**7**	
Unstabilized	93	77	57	51	45	29			Gt 5
1 % HALS-3 2 % BTZ-7	92	75	72	72	69	66	60	58	Gt 0
1 % HALS-2 2 % BTZ-7	93	74	73	73	68	62	55	51	Gt 0
1 % HALS-4 2 % BTZ-7	93	75	74	74	69	63	54	49	Gt 0

clear coat: TSA, basecoat: silver metallic, bake: 130 °C, 30'

1 According to DIN 53151 after additional exposure F; tape test

Table 4.21: Influence of UV absorber and HALS on the glass transition temperature T_g of an acid catalyzed HS-TSA clear coat

Stabilization	**Increase in T_g [°C]**
Unstabilized	24
2 % BTZ-8	9
2 % BTZ-8 1 % HALS-3	5

clear coat: free film, bake: 120 °C, 30', exposure: 500 h exposure C

Figures 4.28 to 4.31 and Table 4.22 illustrate the aging of various clear coats based on different crosslinking mechanisms and under various weathering conditions.

Figure 4.28 shows the time taken for crack development in a one-pack clear PU coat applied to a water-borne, light green metallic basecoat under weathering conditions C. For films that did not crack, the 20° gloss was determined.

The unstabilized clear coat developed cracks already after 2800 hours, that stabilized with 2 % BP-1 and 1 % HALS-3 only after 6800 hours, similar to the clear coat containing only 1 % HALS-3. Obviously, BP-1 provides no protection against cracking in this system. In stark contrast, coatings containing BTZ-8 and HPT-1, each combined with HALS-3 exhibited no cracks after 8000 hours' exposure C. Nor were there any signs of cracking in a coating containing the UV absorber blend of BTZ-8 and HPT-1. The combination of this blend with HALS-3 provides even better gloss retention than when combined with a single UV absorber.

Figure 4.29 shows the effect of 54 months weathering in Allunga, Australia on an acrylic/melamine clear coat applied to a light blue metallic basecoat. The unstabilized sample showed a very considerable reduction in gloss, strong shrinkage but no cracking. The 20° gloss of the stabilized coatings lies between 58 and 68. More significant differences are seen in the colour and the adhesion of the clear coat to the basecoat. Stabilization with only HALS-3 leads to poor colour retention and loss of adhesion, a result of the absence of the UV absorber (see Chapters 3.2.1 and 3.3.2). HALS can prevent cracking and maintain gloss but it cannot protect the coloured basecoat against fading or the degradation of the interface between the basecoat and the clear coat, both caused by UV light. Upon addition of an UV absorber, both these properties improve significantly. There are, however, differences between the UV absorbers. The change in colour ΔE decreases in the order from oxanilide to benzotriazole to triazine. The adhesion increases in the same order (after additional exposure F) and the combination of 1.5 % HPT-1 1 % HALS-3 performs best.

Table 4.22 shows the outdoor weathering results for a 2P-PU clear coat based on an isocyanate-crosslinked acrylic polyol on the same basecoat as in Figure 4.29 after 7 years' exposure in Allunga. The gloss retention of all coatings including the unstabilized control is similar over the first 2 years. Thereafter, the unstabilized coating begins to break down. Both stabilized coatings perform similarly up to 4 years, whereupon that with 1.5 % BTZ-8/1 % HALS-2 undergoes a higher loss of gloss compared to that with 1.5 % HPT-1/1 % HALS-2. Adhesion tests after 5.5 years revealed great differences between the UV

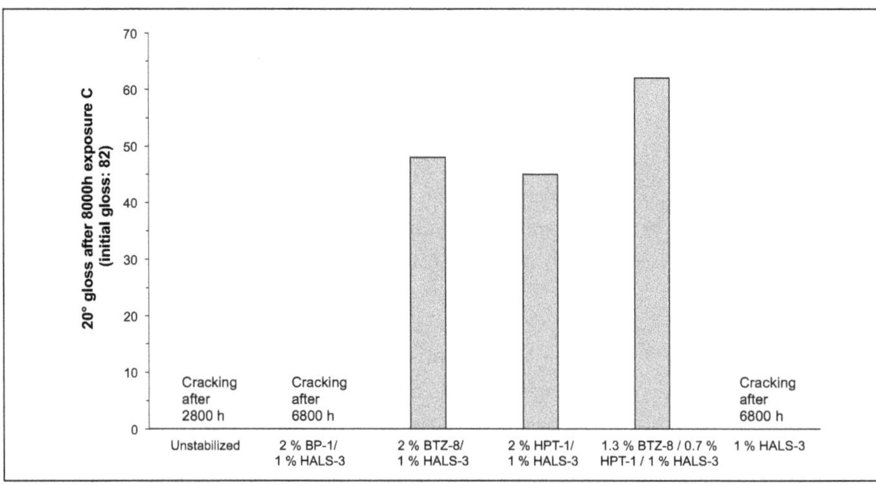

Figure 4.28: Accelerated weathering of a 1P-PU clear coat (exposure C)
basecoat: water-borne, light green metllic, bake: 140 ℃, 20'

Table 4.22: 7 years Allunga exposure of a 2P-PU clear coat

Stabilization	**20° gloss after ... years in Allunga (5°N, bb, unheated)**								**Adhesion[1] after 5.5 years exposure**	
	0	**1**	**2**	**3**	**4**	**5**	**6**	**7**	**before** additional exposure F	**after** additional exposure F
Unstabilized	91	87	80	67	47	15[2]			-	-
1.5 % BTZ-8 1 % HALS-2	90	88	86	82	76	60			Gt 5	Gt 5
1.5 % HPT-1 1 % HALS-2	90	88	86	85	75	67	67	65	Gt 0	Gt 0

basecoat: light blue metallic bake: 80 °C, 30'

1 According to DIN 53151; tape test 2 Cracking after 4.5 years

absorber/HALS blends. Total loss of adhesion was observed with BTZ-8 and HALS-2 (Gt 5), before and after additional exposure F, whereas the HPT-1/HALS-2 combination completely preserved adhesion (Gt 0).

An alternative to 2P-PU coatings are those based on epoxy resins. 2P-epoxy coatings often contain bases serving as curing catalysts. In Chapter 4.2.2.5, the risk of yellow discolouration

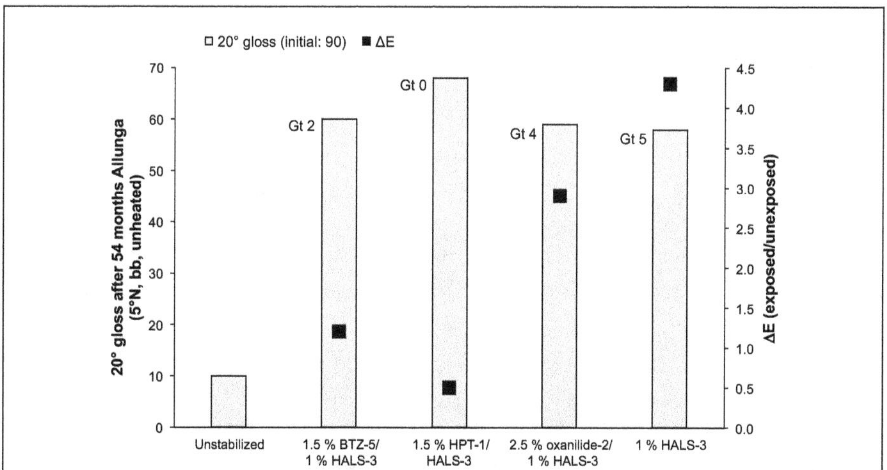

Figure 4.29: Gloss, colour retention and adhesion of an acrylic/melamine clear coat applied to a light blue metallic basecoat after 54 months' exposure in Allunga/Australia
bake: 130 °C, 30'

through the reaction of phenolic UV absorbers such as benzophenones and benzotriazoles with the basic catalyst was discussed. Hitherto, oxanilide UV absorbers were used in such formulations even though hydroxyphenyl-s-triazines, another class of phenolic UV absorbers, do not suffer from this same disadvantage [37].

Figure 4.30 shows the gloss retention of a 2P-epoxy clear coat based on carboxyl- and epoxy-functionalized acrylates cured in the presence of a tertiary amine. After more than

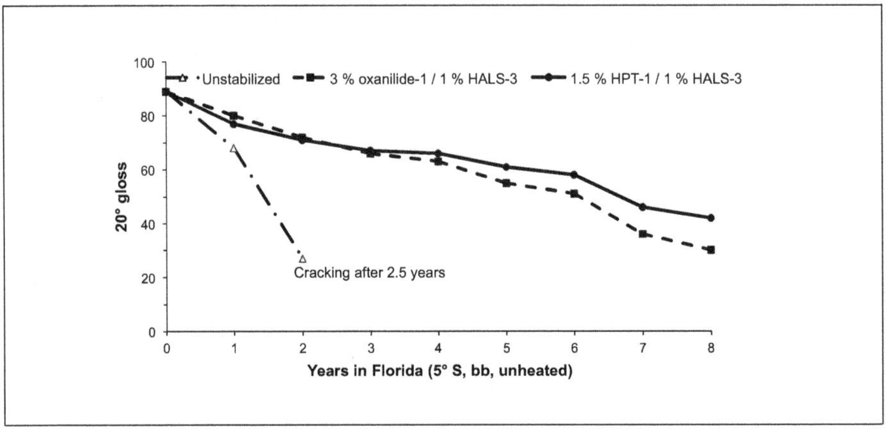

Figure 4.30: Stabilization of a 2P-epoxy clear coat
basecoat: silver metallic bake: 80 °C, 30'

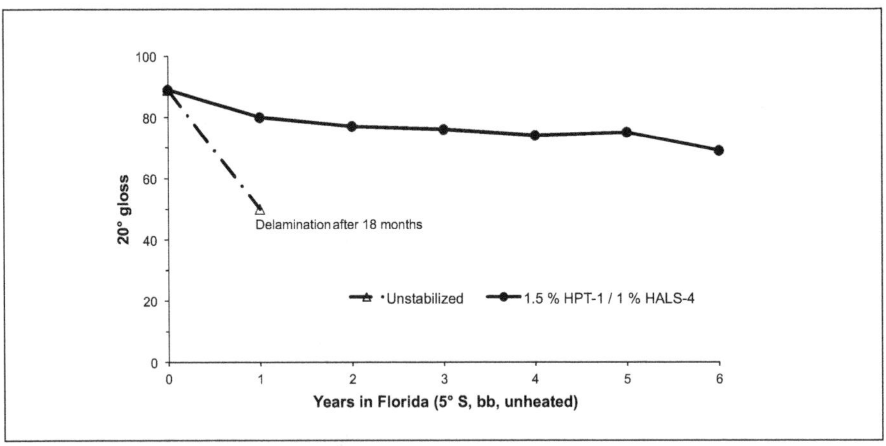

Figure 4.31: Stabilization of a metal catalyzed alkoxy silane based clear coat
clear coat: epoxy silanol curing acrylic [120, 121], catalyst: tris-acetylacetonate, basecoat: white, bake: 140 °C, 30'

8 years of exposure in Florida, the coating stabilized with 1.5 % HPT-1 has even higher gloss than that containing twice as much oxanilide-1 (both UV absorbers were combined with HALS-3). The unstabilized clear coat suffered relatively strong gloss deterioration right from the start and developed cracks after 2½ years.

In addition to their unwelcome reaction with certain amines, benzophenones and benzotriazoles also tend to react with certain metal compounds such as aluminium tris-acetylacetonate, used as crosslinking catalyst (see Chapter 4.2.2.5). Here again, only hydroxyphenyl-s-triazines find application (see Figure 4.31).

The gloss of the unstabilized clear coat/white solid shade basecoat began to decrease at the onset of the weathering test and complete delamination was observed after 18 months. One reason for this rather fast degradation could be the limited hydrolytic stability of the Si-O-C bond [137]. On the other hand, the same coating system formulated with HPT-1 and HALS-4 had not cracked after 6 years in Florida and still exhibited a gloss of approximately 70 units.

4.2.4.3 Results for water-borne clear coats

In the early fifties, work was already underway to develop water-based binders with the goal of replacing solvent-based paints [167]. Water-borne basecoats have been in regular use since the mid-eighties [168] and in 1992, the first water-borne clear coat was introduced in the automotive industry (Opel in Eisenach/Germany) [169].

During the early development of water-borne clear coats, one of the questions that had to be answered was if such modified binders have weathering properties comparable to

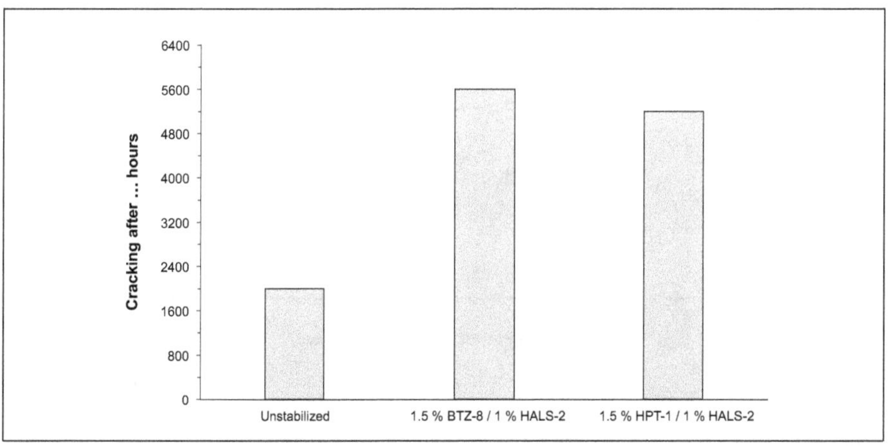

Figure 4.32: Cracking of a water-borne acrylic/melamine clear coat (exposure E)
basecoat: water-borne, white, bake: 80 °C, 10' + 140 °C, 20'

those of solvent-based binders. This is clearly the case for both for OEM and refinish systems as shown in the following examples (Figs. 4.32-4.36).

Figure 4.32 shows the accelerated weathering results (exposure E) for a water-borne clear coat applied to a white water-borne basecoat. The unstabilized clear coat cracked after 2000 hours' weathering whereas the stabilized counterparts, with either 1.5 % BTZ-8 or 1.5 % HPT-1 and 1 % HALS-2, withstood weathering for 5200 and 5600 hours, respectively.

Figure 4.33 shows the effects of weathering of a water-borne clear coat applied on top of a water-borne, dark blue metallic basecoat. The gloss of a sample stabilized with HPT-1/ HALS-2 was acceptable after 6 years' exterior exposure. After 3 years' exposure, this same sample displayed excellent adhesion (Gt 0), whereas the unstabilized sample suffered complete delamination (Gt 5) after 3.5 years. Colour retention also varied considerably. The colour change ΔE of the stabilized system was 0.4, without stabilization, 3.5.

Figure 4.34 shows that water-borne clear coats are as weather resistant as their solvent-based counterparts. A clear coat stabilized with 1.5 % HPT-1 and 1 % HALS-3 shows very good gloss retention (20° gloss approx. 60 units) even after 6 years' weathering in Florida.

An important milestone in the evolution of water-borne formulations was the development of water-based basecoats for car refinishing [170] and the corresponding refinish clear

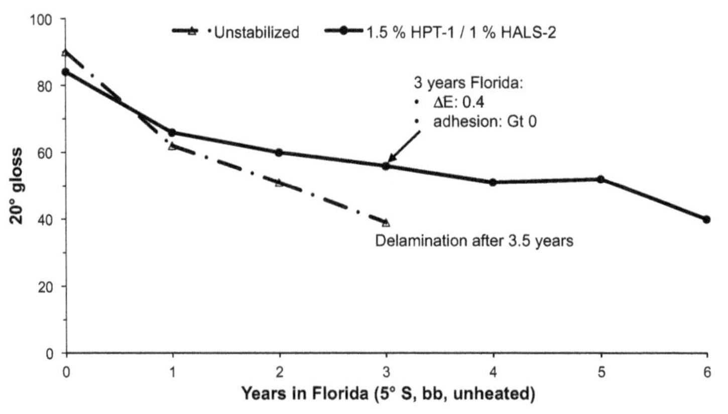

Figure 4.33: Gloss profile of a water-borne clear coat/basecoat system
clear coat: water-borne acrylic/melamine, basecoat: water-borne, dark blue metallic, bake: 80 °C, 10' + 140 °C, 20'

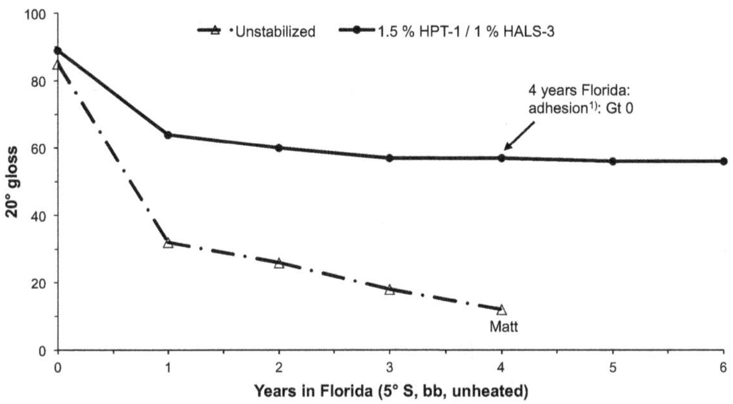

Figure 4.34: Gloss profile of a water-borne clear coat/basecoat system
clear coat: water-borne acrylic/melamine, basecoat: water-borne, silver metallic, bake: 70 °C, 20' + 140 °C, 30'

1 According to DIN 53151 after additional exposure F; tape test

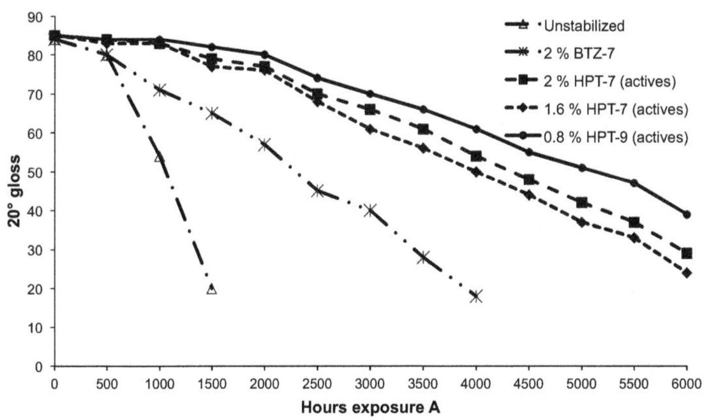

Figure 4.35: Gloss retention of a water-borne 2P-PU refinish clear coat on a black water-borne basecoat
cure clear coat: 30' flash off + 60 °C, 30'; DFT: 50 µm, stabilization: UV absorber + 1 % HALS-2

coats. Figure 4.35 shows the weathering results (Xenon WOM exposure, exposure A) of a water-borne 2P-PU Refinish clear coat applied to a black water-borne basecoat. The best stabilization is achieved with a combination of HPT-9 and HALS-2, the combination of HPT-7 and HALS-2 is only slightly inferior.

Figure 4.36 shows the colour retention of a 2P-PU refinish clear coat applied over a white water-borne basecoat after Xenon WOM exposure (exposure B). The best colour retention is achieved with HPT-9 or HPT-7, in combination with HALS-2.

Even though the data clearly indicates that water-borne clear coats represent a valid alternative to traditional solvent-based systems, they have not been broadly adopted in automotive OEM or refinishing.

4.2.4.4 Results for powder clear coats

The evolution of the car coating technology from medium to high-solids and finally water-based clear coats has made a major contribution towards the reduction of solvent emissions. The development of automotive powder coatings in the early nineties [171–173] was fuelled by the desire to further reduce or even eliminate solvents from the coating process altogether. Although this technology was still practiced at some car manufactures in the first decade of the 21 century, it is only of minor importance today. The coatings were

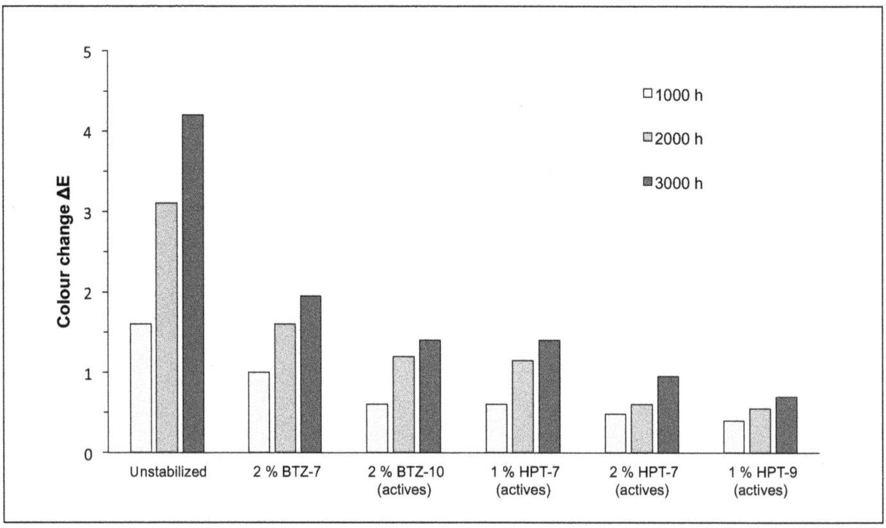

Figure 4.36: Colour retention of a water-borne 2P-PU refinish clear coat over a white water-borne basecoat after weathering
cure clear coat: 30' flash off + 60 °C, 30'; DFT 40 µm, stabilization: UV absorber + 1 % HALS-2; exposure: Xenon WOM (exposure B)

based mainly on carboxyl and anhydride-functional acrylic resins cured with epoxy resins. In addition to such completely solvent-free clear powder coatings, water-borne powder slurries found some use due to their easy handling [174]. Nevertheless, it is still interesting to examine the stabilization of powder systems and the selection criteria for the optimal stabilizer package.

The use of powder coatings in the car industry was not new at this time. Prior to the introduction of powder coatings, powdered fillers were already used [175]. In order to achieve the levelling (appearance) of solvent-based or water-borne clear coats, however, clear powder coatings had to be applied at higher film builds – approx. 60 to 80 µm.

The development of powder coatings prompted the development of new light stabilizers adapted to the requirements of this new technology. Whereas priority in the early eighties was set on liquid light stabilizers for use in high-solids (= low-solvent) or water-borne formulations, powder coatings required *solid* light stabilizers for reasons of convenience and ease of processing. The melting point of additives used for powder coatings must be in a specific range, not too high so to avoid mixing problems during extrusion and not too low so that the glass transition temperature and storage stability of the powder is negatively impacted. Additives melting in the range as the extrusion temperature, 70 to 110 °C, appear to be most suitable. Examples include BTZ-3 (despite its melting point of

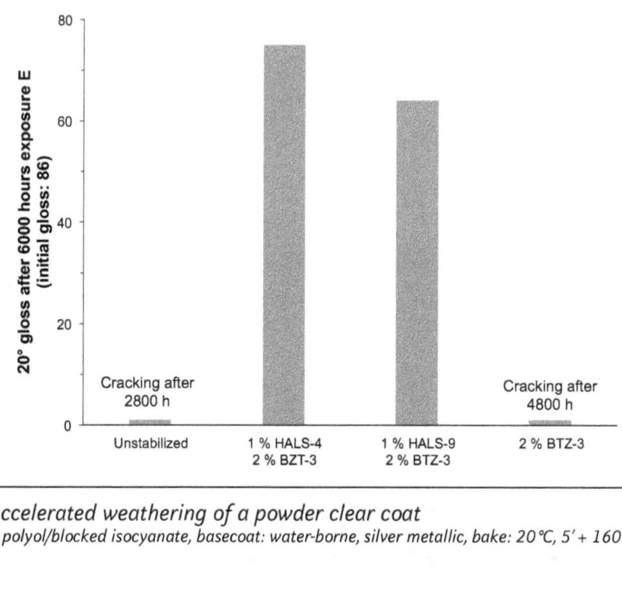

Figure 4.37: Accelerated weathering of a powder clear coat
clear coat: acrylic polyol/blocked isocyanate, basecoat: water-borne, silver metallic, bake: 20 °C, 5' + 160 °C, 20'

>130 °C), BTZ-4, HPT-2, HPT-3, HPT-4, HALS-4 (again, despite its melting point of >140 °C) and HALS-9. Additives with high melting points can be used if they dissolve readily in the resins of the formulation.

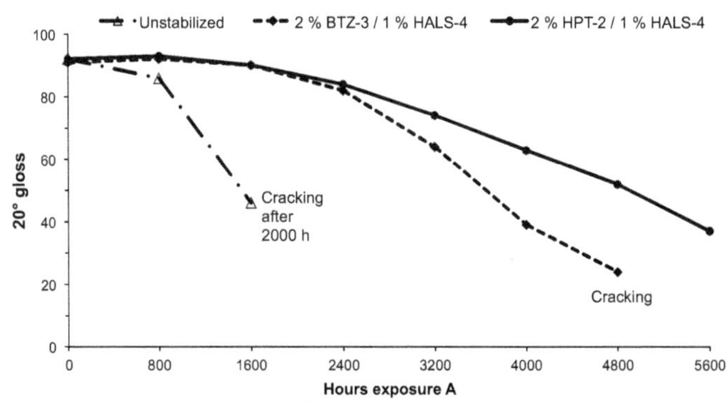

Figure 4.38: Accelerated weathering of a powder clear coat
clear coat: carboxy-/epoxy functional acrylate, basecoat: water-borne, silver metallic, bake: 145 °C, 30'

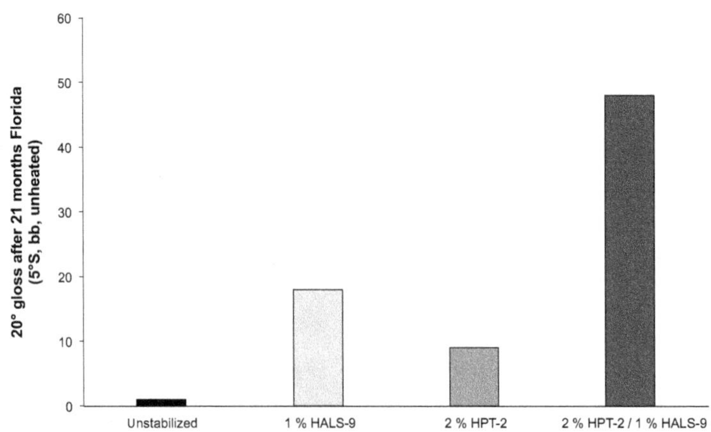

Figure 4.39: Outdoor exposure of a powder clear coat
clear coat: epoxy-/anhydride functional acrylate, basecoat: water-borne, silver metallic, bake: 140 °C, 30', initial gloss: 90°

Figure 4.37 concerns the weathering of a polyurethane based powder clear coat applied to a water-borne basecoat. Very good gloss retention was observed for all coatings stabilized with BTZ-3 and HALS after 6000 hours (weathering condition E). The coating without light stabilizer developed cracks after 2800 hours, whilst another containing only BTZ-3 resisted cracking for 4800 hours.

Figure 4.38 shows gloss values of a clear coat based on a carboxy-/epoxy functional acrylate applied to a water-borne, silver metallic basecoat. The coatings stabilized with BTZ-3 and HALS-4 or with HPT-2 and HALS-4 exhibit a far better stability than that without any light stabilizers, which cracked after 2000 hours weathering (conditions A).

Figure 4.39 shows coatings based on an epoxy-/anhydride functional acrylate after 21 months' exposure in Florida. The system itself is not very weather resistant but the use of light stabilizers has a significant impact, especially the combination of 2 % HPT-2/1 % HALS-9.

4.2.4.5 Results for UV-curable clear coats

UV-cured clear coats, like high-solids, water-borne and powder clear coats which were discussed in previous chapters, can be considered environmentally friendly. Due to the misconception that UV-activated photoinitiators and UV absorbers are not compatible due to overlapping absorption spectra, UV-curing paints were also deemed unsuitable for outdoor use.

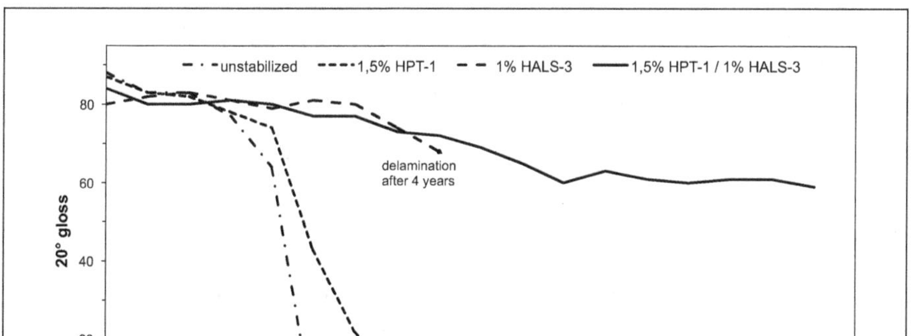

Figure 4.40: Outdoor exposure of a UV-curable clear coat
clear coat: aliphatic polyester triurethanetriacrylate/hexanediol diacrylate (60/40), DFT: 40 µm, PI: 3 % PI-II, curing 2 x 80 W/cm, 2 x 10 m/min, basecoat: water-borne, dark blue metallic (pre-baked)

Table 4.23: Accelerated weathering of a UV-curable clear coat

Stabilization	Cracking after ... hours exposure A	Color change ΔE after ... hours exposure A		
		3200	6400	9600
Unstabilized	2800	-	-	-
1.5 % HPT-1	3600	0.2	-	-
1 % HALS-3	8400	1.0	2.0	-
1.5 % HPT-1 1 % HALS-3	> 10.000	0.3	0.3	0.3

clear coat: aliphatic polyester triurethanetriacrylate/hexanediol diacrylate (60/40), 40 µm,
PI: 3 % PI-II, curing 2 x 80 W/cm, 2 x 10 m/min
basecoat: water-borne, dark blue metallic (pre-baked)

The issues concerning the combined use of UV absorbers with photoinitiators were discussed in detail in Chapter 4.2.2.5. The UV absorber and photoinitiator must be carefully selected so that sufficient light remains for the excitation of the photoinitiator. Hydroxyphenyl-s-triazine UV absorbers, in particular, prove to be particularly suitable for such formulations.

The weathering of such clear coats will now be analysed in model formulations. Since these formulations generally contain reactive diluents but no solvents, the use liquid UV absorbers and HALS, which can readily miscible, is warranted. Table 4.23 shows the weathering results (conditions A) of a UV-curing clear coat containing PI II applied to a waterborne, dark blue metallic basecoat.

The time when first signs of cracking appeared as well as the colour change, ΔE, are comparable to other formulations. Resistance to cracking and colour change, however, were much improved by the use of suitable light stabilizers. Coatings stabilized with the combination of HPT-1 and HALS-3 were intact even after 10,000 hours of weathering (condition A) with very good gloss (20° gloss ~65 units) and colour retention (ΔE=0.3).

The findings of the accelerated weathering test matched well with those from 8.5 years' outdoor exposure in Florida (see Figure 4.40). HPT-1 or HALS-3 alone provide some stabilization but complete delamination occurs nonetheless after 4 years' exposure. In the case of HALS-3, high gloss retention was observed. As expected from the accelerated weathering, the combination of HPT-1 with HALS-3 guarantees the integrity of the coating for 8.5 years and more with very good gloss retention (~60 units).

Similar outdoor weathering tests were carried out on a two-coat finish on a water-borne, red solid shade basecoat (see Figure 4.41). Here, the previous results were confirmed. The combination of UV absorber and HALS in this case with HALS-2, provided satisfying gloss retention after 8.5 years weathering in Florida without any sign of cracking or delamination.

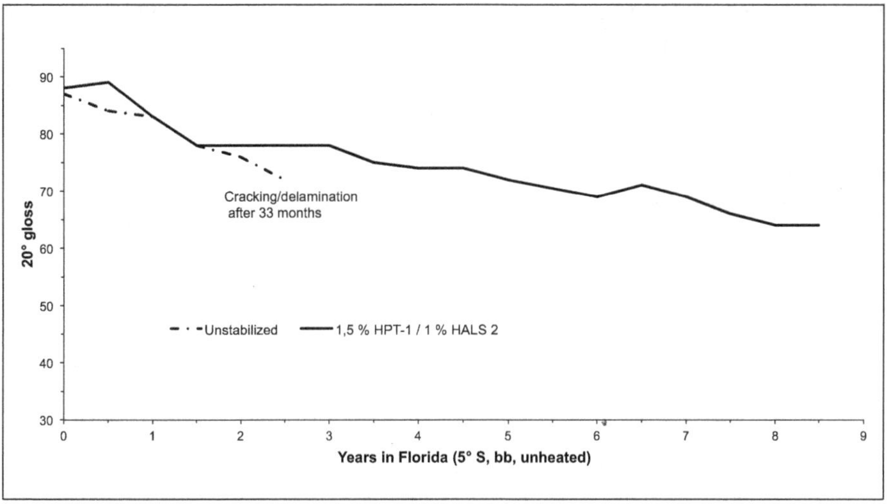

Figure 4.41: Outdoor exposure of a UV-curable clear coat
clear coat: aliphatic polyester triurethanetriacrylate/hexanediol diacrylate (60/40), DFT: 40µm, PI: 3 % PI-II, curing 2 x 80 W/cm, 2 x 10 m/min, basecoat: water-borne, red solid shade (pre-baked)

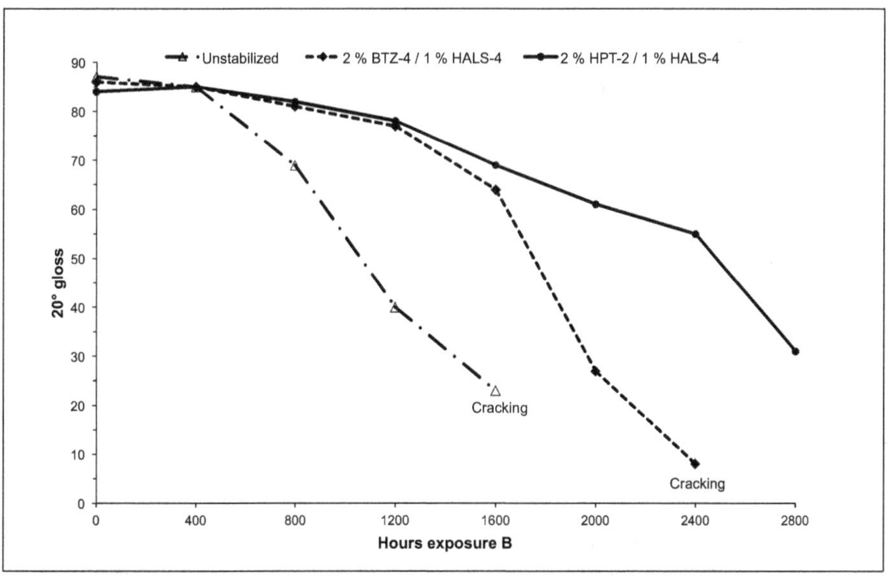

Figure 4.42: Accelerated weathering of a UV-curable powder clear coat
clear coat: unsaturated polyester/urethane acrylate (70/30), 65 µm, PI: 2.25 % PI-II/0.75 % PI-III, curing 2 x 80 W/cm, 2 x 8 m/min, basecoat: water-borne light blue metallic (pre-baked)

The unstabilized clear coat developed cracks after 33 months and delamination of the clear coat from the basecoat was evident.

The tests discussed above were made with solvent-free, liquid, UV-curing clear coats, which are generally blends of a high viscosity oligomer with a low-viscosity monomer having the optimal consistency for easy application. Other UV-curable coating technologies, water-based and, the latest development in this field, clear powder coatings, were originally developed for interior applications on wood, for example MDF (medium-density fibreboard) [176, 177], but could also be modified for use in automotive finishes [145].

Figure 4.42 illustrates an automotive weathering test carried out on a UV-curable clear powder coating based on an unsaturated polyester urethane acrylate, originally developed for wood varnishes, applied on a water-borne, light blue metallic basecoat. A mixture of the photoinitiators PI-II and PIII was used to cure the topcoat. Note that, similar for thermosetting powder coatings (Chapter 4.2.4.4), solid UV absorbers and HALS are preferred for the production of the powder coatings by extrusion.

The clear coat without light stabilizer exhibited cracking after 1600 hours of weathering (condition B). The combination of 2 % BTZ-4 and 1 % HALS-4 significantly improved the weather resistance, extending the time before cracking to 2400 hours. No cracking was observed after 2800 for a coating stabilized with 2 % HPT-2 and 1 % HALS-4.

Chapters 4.2.4.2 to 4.2.4.5 show conclusively that clear coats containing blends of UV absorbers and HALS withstand weathering over long periods of time, irrespective of their nature: solvent-based, water-based, powder, or liquid and solvent-free. If the right light stabilizer is used, the underlying substrate (for example basecoat) is also protected. The most suitable UV absorbers both for liquid and powder coatings are benzotriazoles and hydroxyphenyl-s-triazines. The choice of stabilizers or combinations thereof depends on the type of coating and the performance requirements. In the case of HALS, the nature of the system dictates whether a basic or non-basic HALS is required.

4.2.4.6 Coatings on plastic substrates

The use of plastics has significantly increased over the years, both in industrial and automotive applications. Industrial applications include packaging, building & construction, roofing, greenhouses, aerospace, household, to name a few. With a focus on weight reduction (less fuel consumption), freedom of design, cost efficiency and recyclability [178], plastic is replacing metal in cars inside and out; examples include bumpers, mirror housings, body panels, headlamps, sunroofs and dashboards.

The coatings used for the outside car parts are expected to be as durable in terms of colour and gloss retention and crack-resistance as those used to paint the steel body. However, coatings for plastics need to be more flexible than those for steel and therefore require other components, which in turn my affect their durability. Furthermore, additives or low molecular weight components from the plastic may migrate into the coating [179] or light stabilizers may migrate in the reverse direction [132, 180–182] leading to a depletion of additives in the topcoat.

In Chapters 4.2.2.3 and 4.2.3.3 the distribution profiles of reactable and non-reactable light stabilizers on plastic and steel substrates before and after Xenon WOM exposure were presented. The results clearly verify the migration of non-reactable light stabilizers into the plastic substrate, whereas reactable light stabilizers are largely retained in the coating. This explains the inferior durability of coatings containing non-reactable light stabilizers applied to plastic substrates such as thermoplastic polyolefines (TPO) as opposed to steel.

Figure 4.43 shows a comparison of steel coatings containing either the reactable UV absorber HPT-1 or the non-reactable UV absorber BTZ-4. As migration into the substrate is not possible in this case, the weathering results are very comparable. The slight advantage of HPT-1 stems from the superior photostability of the hydroxyl-phenyl-triazine (see Chapter 4.4.1).

The same comparison but on a TPO substrate (thermoplastic polyolefin) paints a different picture (Figure 4.44). Here, the performance of the reactable UV absorber HPT-1 is clearly superior since, in contrast to non-reactable BTZ-4, HPT-1 does not migrate into the plastic substrate.

A further example of the advantages of reactable stabilizers is shown in Figure 4.45. The reactable HALS-10 displays superior performance vs. the non-reactable HALS-3 whether

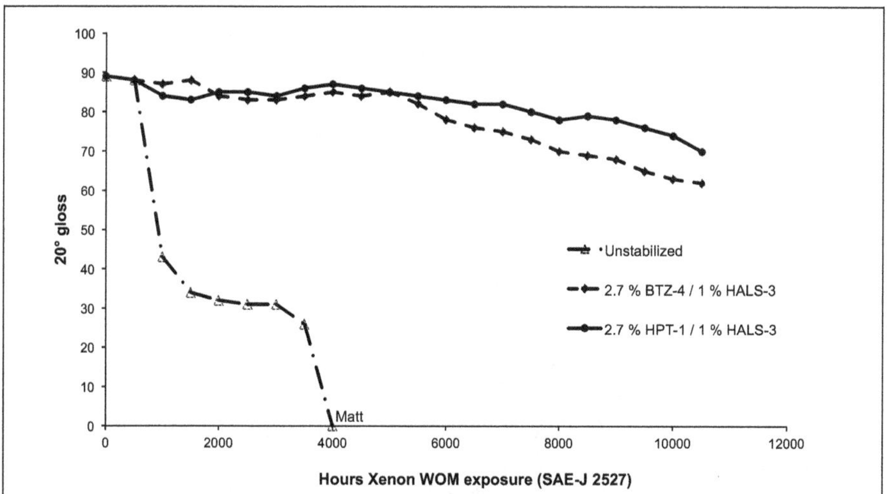

Figure 4.43: Comparative results of reactable (HPT-1) vs non-reactable (BTZ-4) UV absorbers in a 2P-PU clear coat on a metallic basecoat, substrate: steel

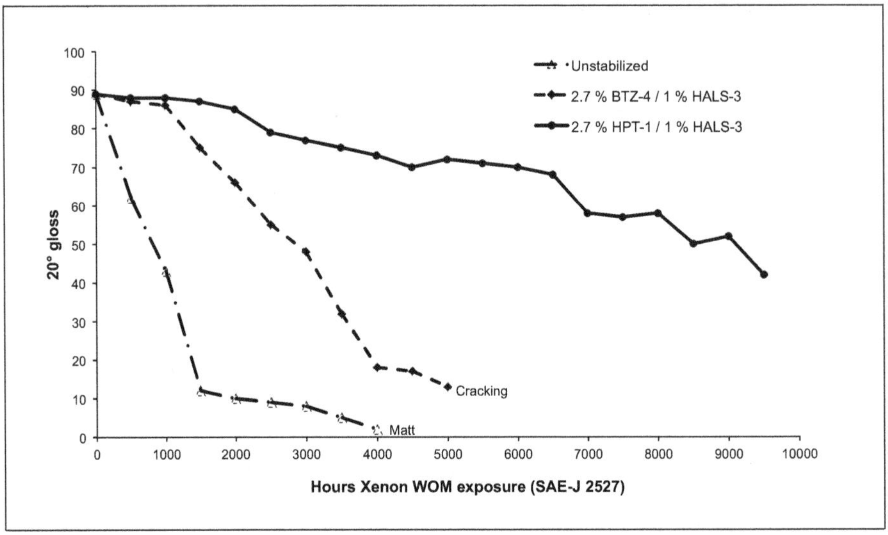

Figure 4.44: Comparative results of reactable (HPT-1) vs non-reactable (BTZ-4) UV absorbers in a 2P-PU clear coat on a metallic basecoat, substrate: TPO

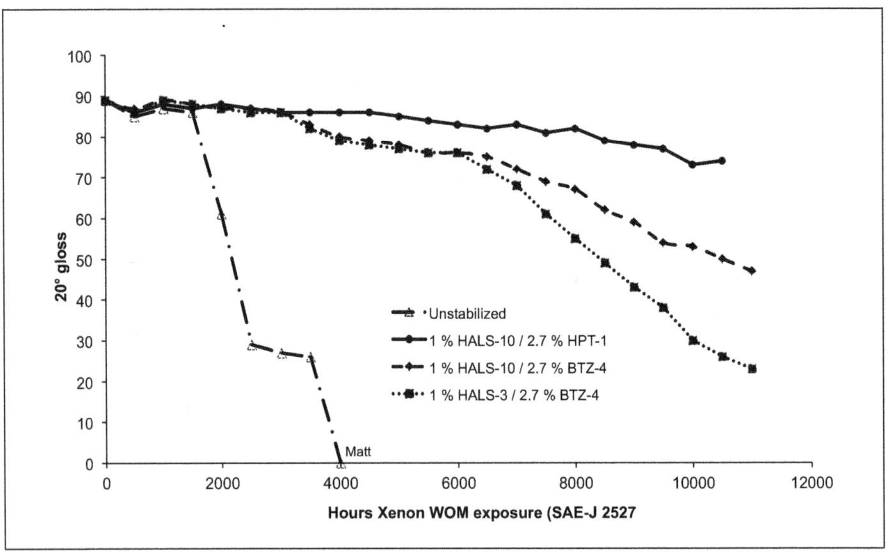

Figure 4.45: Comparative results of reactable (HALS-10) vs non-reactable (HALS-3) HALS in a 2P-PU clear coat on a metallic basecoat, substrate: TPO

in combination with the BTZ-4 or HPT-1. The combination of reactable UV absorber and HALS, HALS-10 and HPT-1, guarantees durability for plastic coatings comparable to that achievable on steel.

Special attention must be paid to the selection of the light stabilizers in the case of polycarbonate substrates. In addition to the migration issues discussed above, the base sensitivity of polycarbonate has to be taken into account. In the presence of traces of water, strongly basic amines cause polycarbonate degradation [183] resulting in delamination. This excludes the use of basic HALS such as HALS-1, HALS-2, HALS-6 or HALS-7. In addition, polycarbonate is particularly sensitive to short wavelength UV light, which is best filtered by hydroxy-phenyl-s-triazine UV absorbers. The optimal stabilizer package is thus one composed of HPT-1 (reactable UV absorber) and HALS-10 (reactable and non-basic HALS). Figure 4.46 shows the performance of this combination in a weathering test (Florida exposure, 4.5 years, 5° S, black box, unheated).

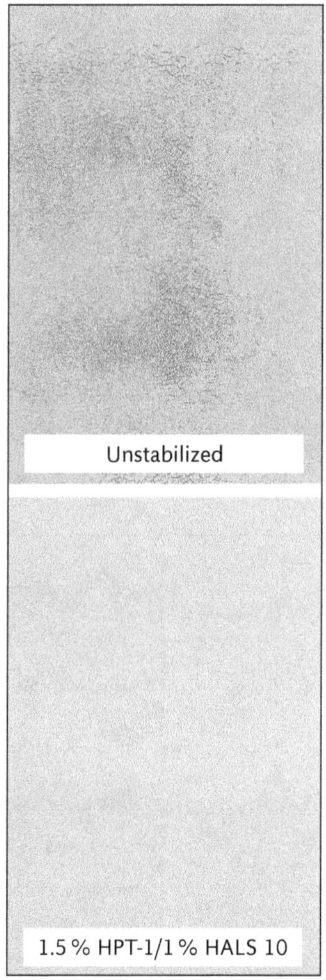

Figure 4.46: Florida exposure of a 2P-PU clear coat over silver metallic basecoat on polycarbonate
exposure: 4.5 years Florida (5° S, bb, unheated)

Headlamps and potentially even side or rear windows, if mechanical requirements such as scratch resistance are met, where transparent polycarbonate can replace glass, are important applications for coatings over plastic substrates. Scratch resistance can be attained by application of plasma coatings (CVD = Chemical Vapour Deposition) [184], siloxane-based hardcoats [184] or thin UV-curable clear coats [185].

Figure 4.47 shows an example of the latter, wherein the UV-curable clear coat had a dry film thickness of 13 µm after curing. The UV absorber concentration of the samples is higher than that seen in previous studies since lower coating thicknesses require higher amounts of the UV absorber to reach the same extinction (filter effect). Thus, whereas it is normal to add between 1.5 and 2 % HPT-1 to a coating with a thickness of 40 µm, 4.5 to 6 % of this additive must be used for a coating of only of 13 µm thickness (see Chapter 3.2.1). It is interesting to note that even at this high concentration,

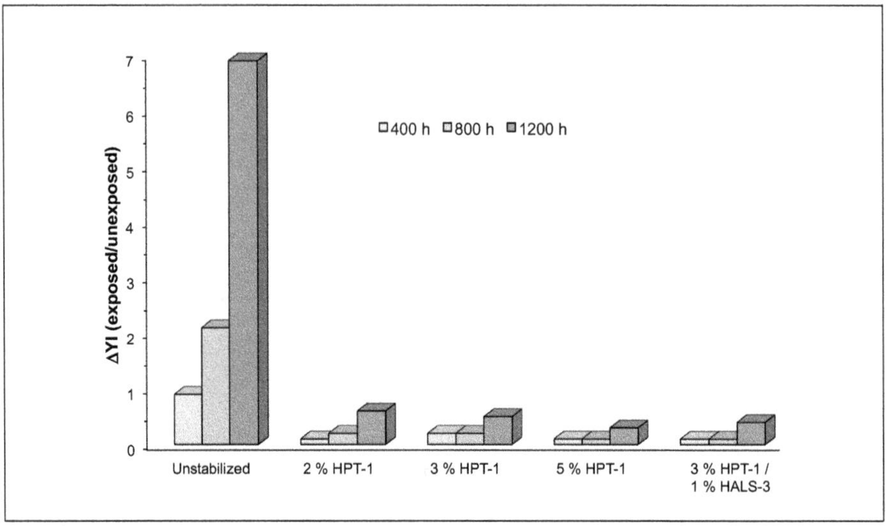

Figure 4.47: Accelerated weathering of a UV-curable clear coat applied on transparent polycarbonate (exposure B)
clear coat: aliphatic urethane acrylate/hexanediol diacrylate (40/60), DFT: 13 µm, PI: 2.25 % PI-I/0.75 % PI-III, curing 2 x 80 W/cm, 2 x 8 m/min

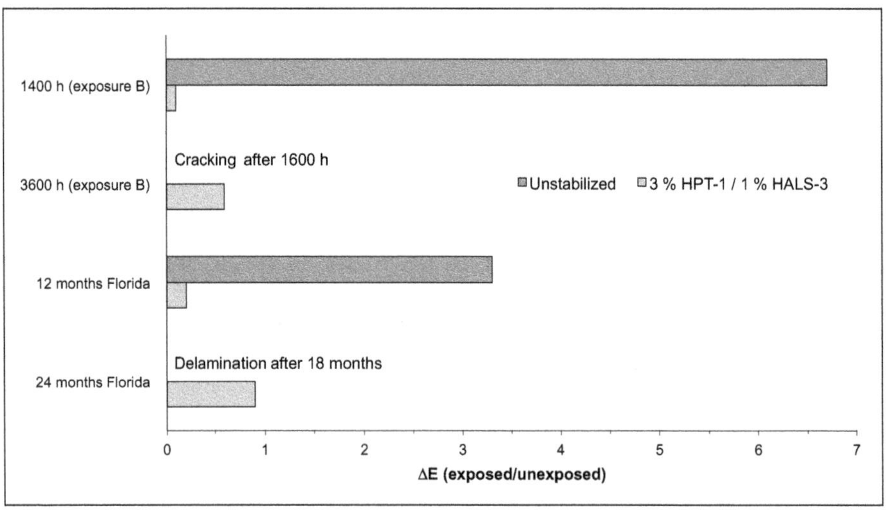

Figure 4.48: Outdoor exposure of a UV-curable clear coat applied on transparent polycarbonate
clear coat: aliphatic epoxyacrylate/aliphatic urethane acrylate/tripropylene glycol diacrylate/trimethylol propane triacrylate (45/35/10/10), DFT: 20 µm, PI: 2.7 % PI-I/0.3 % PI-IV, curing 2 x 80 W/cm, 2 x 10 m/min, exposure: exposure B, Florida (45° S, bb, unheated)

the UV absorber had no adverse effect on the UV curing of the clear coats when a combination of PI-I and PI-III were used. The effect of 1200 hours of weathering (conditions B) on the samples was quantified as change in the yellowness index, ΔYI.

Lowest yellowing was observed for samples containing 5 % HPT-1 and became more evident over time due to the improved filter effect at this high UV absorber concentration. Very good results were also achieved with a combination of 3 % HPT-1 and 1 % HALS-3, as well as combinations of HPT-1 with other non-basic HALS such as HALS-10 or HALS-11. HALS is indispensable for the long-term stability of the paint because it protects the clear coat against cracking.

Another UV-curable clear coat applied on transparent polycarbonate is shown in Figure 4.48. Here, the dry film thickness was 20 to 22 µm requiring the use of 3 % HPT-1 (corresponds to ~1.5 % HPT-1 in a 40 µm film). The colour change ΔE was determined after 3600 h accelerated weathering (condition B) as well as 24 months' outdoor exposure in Florida (45° S, bb, unheated). The coating stabilized with 3 % HPT-1/1 % HALS-3 shows significantly less colour change than the unstabilized control and, more importantly, did not crack.

4.2.4.7 UV protection of epoxy-based fibre reinforced plastics

Composite materials are prevalent in a number of applications including automotive, hang-on parts for motor bikes, aerospace, marine, civil engineering and sporting goods such as bicycles. Of the fibre-reinforced materials, carbon fibre reinforced plastics (CFRP) are often employed since they possess the high stiffness and strength of a metal without the weight [186].

Suitable polymers carriers include polyesters, vinyl esters, epoxy resins, polyimides or polyamides. Because of their relatively low cost and high performance, amine-cured epoxies (for example, polyetheramines) [187, 188] are the polymer matrices of choice for many of the above applications.

CFRP based on aromatic epoxies are finding use as design elements for automobiles and other applications (substrate coated with clear coat only). As these composites are inherently light sensitive, unless protected by pigmented coatings, their stabilization in such applications is very challenging. Fundamental studies conducted with cut-off filters revealed that these materials are highly sensitive at wavelengths both in the UV (for example 280 to 380 nm) and around 400 nm [189].

The spectral sensitivity of epoxy-based CFRP was determined using cut-off filters (see Figure 4.49). Substrates coated with a 2P-PU clear coat stabilized uniquely with HALS were subjected to 4000 h Xenon WOM irradiation followed by exposure to humidity (96 h, 40 °C at 98 % r.h.). The results clearly show that protection at wavelengths up to and beyond 400 nm is essential to prevent light induced degradation of the underlying epoxy matrix.

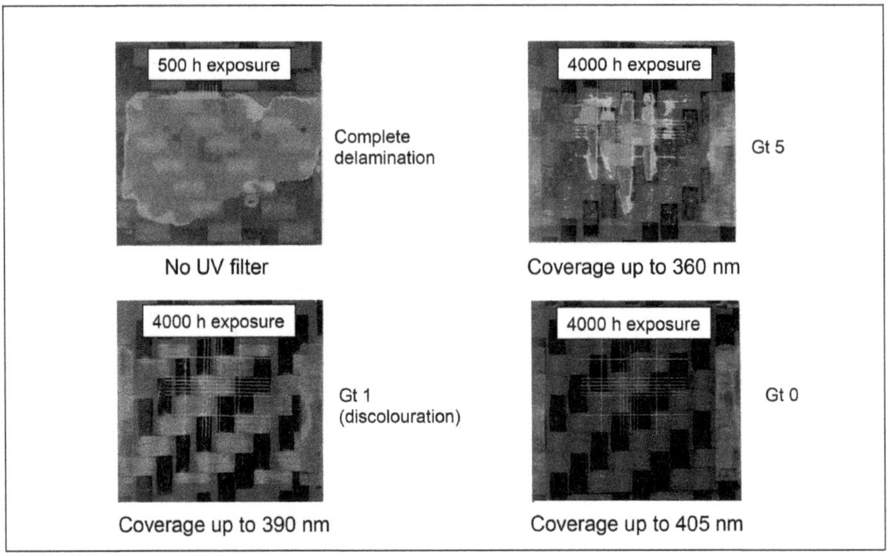

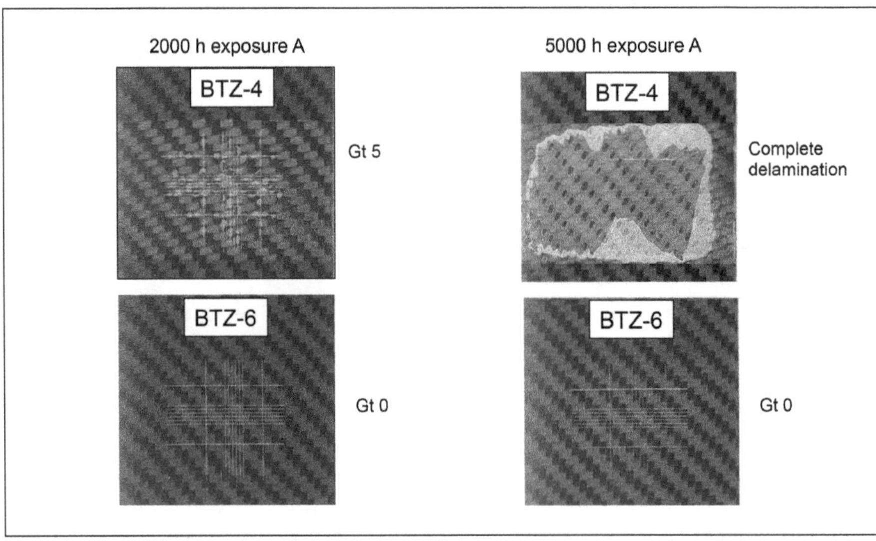

Figure 4.49: Determination of the spectral sensitivity of epoxy based CFRP using cut-off filters
clear coat: 2P-PU (stabilized with 1 % HALS-2 only), substrate: CFRP, exposure: 4000 h Xenon WOM, adhesion test according to DIN 53151 after additional exposure F; tape test

Figure 4.50: Comparison of a conventional (BTZ-4) and a red-shifted (BTZ-6) UV absorber
clear coat 2P-PU (contains 1 % HALS-2), DFT: 2 x 60 µm, substrate: CFRP

Therefore, the protection of such substrates in clear coat applications requires UV absorbers with pronounced absorption into the near visible with sufficiently high extinction around 400 nm and superior photostability with minimal inherent colour (red-shifted UV absorbers) [189]. These requirements are met with BTZ-6.

Comparative experiments in clear coats over CFRP using BTZ-6 and the conventional UV absorber BTZ-4 (both in combination with HALS-2, see Figure 4.50) confirm that the spectral coverage provided by the latter is not broad enough to protect the substrate: delamination of the clear coat occurs after only 2000 h Xenon WOM exposure. In contrast, the coating with BTZ-6 showed no signs of delamination or discoloration even after 5000 h Xenon WOM exposure.

Similar results were obtained after two years' exposure in Florida (5° South, bb, unheated). The analysis of the weathered coatings was made subsequent to exposure to 98 % relative humidity at 40 °C through a tape test of a cross cut (Figure 4.51). Again, the coating with the conventional UV absorber BTZ-4 suffered complete delamination as evidenced by the severe whitening due to penetration of water into the coating. The coating containing BTZ-6 was completely unaffected.

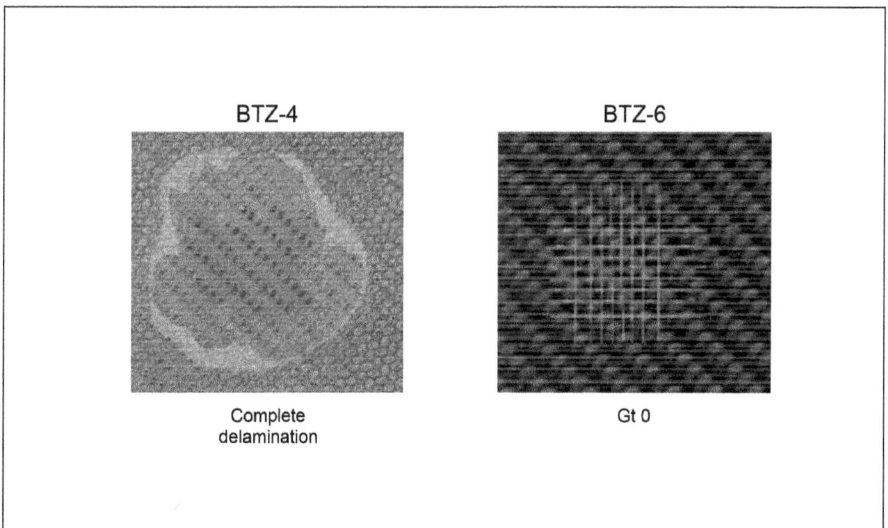

Figure 4.51: Comparison of the conventional and red-shifted UV absorbers BTZ-4 and BTZ-6 in a Florida outdoor exposure test

clear coat: 2P-PU (contains 1 % HALS-2), DFT: 2 x 60 µm, substrate: CFRP, exposure: 2 years Florida, 5° S, bb, unheated, adhesion test according to DIN 53151 after additional exposure F; tape test

4.2.4.8 Effect of additional basecoat stabilization

In 2014 globally more than 260,000 tons of liquid basecoat formulations were produced for the vehicle industry (passenger cars), 48 % solvent-based and 52 % water-borne [101]. The question often arises if the addition of additional light stabilizer to the basecoat will provide increased weathering resistance. There is no clear answer to this question since the effect of additional stabilizer depends upon the nature of the binder, the pigment content and the other additives present. It has been shown, however, that additional light stabilizer does improve the light and weather resistance of binders and pigments in some cases

The examples in Figure 4.52 and in Tables 4.24 to 4.26 serve to illustrate the effect of additional light stabilizer on basecoat stability.

Figure 4.52 illustrates the behaviour. Water-borne clear coat/water-borne, dark blue metallic basecoat systems aged for four years show gloss values between 37 and 60 units, however, those exhibiting higher gloss contain additional stabilizer in the basecoat. The distinctness of the reflected image (DOI) is very sensitive to coating degradation and can better discriminate between the aged coatings. Those coating systems where only the clear coat is stabilized show a DOI of ~1. Basecoats with an additional 1 % of HALS-2, particularly in combination with the clear coat containing 1.5 % HPT-1/1 % HALS-2, provide a significantly higher DOI.

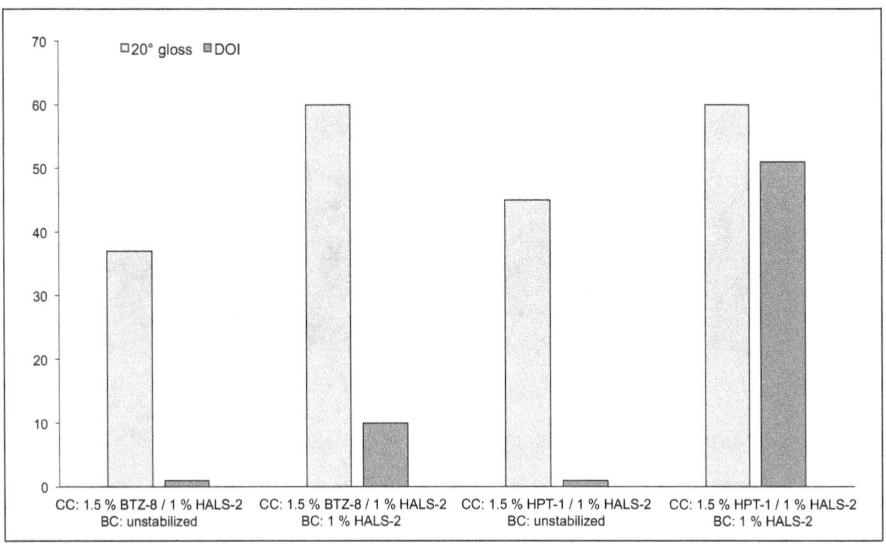

Figure 4.52: Gloss and distinctness of reflected image (DOI) of a water-borne clear coat/ basecoat system

clear coat (CC): water-borne acrylic/melamine, basecoat (BC): water-borne, dark blue metallic, bake: 80 °C, 10' + 140 °C, 20', initial gloss: 92; initial DOI: 90

Table 4.24: Effect of additional basecoat stabilization on the properties of a HS-2P-PU clear coat

Stabilization	Basecoat: silver metallic (15 µm)		Basecoat: red solid shade (30 µm)	
	20 ° gloss	adhesion[1]	20 ° gloss	adhesion[1]
Clear coat: unstabilized Basecoat: unstabilized	total delamination; no adhesion test possible		total delamination; no adhesion test possible	
Clear coat: 2 % BTZ-7/1 % HALS-2 Basecoat: unstabilized	65	Gt 0	26	Gt 5
Clear coat: 2 % BTZ-7/1 % HALS-2 Basecoat: 1 % HALS-2	82	Gt 0	79	Gt 0

Bake: 140 °C, 20′ DFT: 45 µm
basecoat: water-borne exposure: 4 years Florida (5° S, bb, unheated)

1 clear coat/basecoat adhesion according to DIN 53151 after additional exposure F; tape test

Table 4.25: Effect of additional basecoat stabilization on adhesion

Stabilization	Adhesion[1] after 2 years Florida
Clear coat: unstabilized basecoat: unstabilized	total delamination; no adhesion test possible
Clear coat: unstabilized basecoat: 1.5 % BTZ-8	total delamination; no adhesion test possible
Clear coat: 1.5 % BTZ-8/1 % HALS-2 basecoat: unstabilized	Gt 2
Clear coat: 1.5 % BTZ-8/1 % HALS-2 basecoat: 1.5 % BTZ-8	Gt 0

clear coat: TSA DFT: 30 µm bake: 130 °C, 30′,
basecoat: Pigment blue 60 (5 %) DFT: 10 µm substrate: cathodic primer

1 tested was the adhesion of clear coat/basecoat system on the cathodic primer. Assessment according to DIN 53151

The binder of the unstabilized high-solids, 2P-PU clear coat applied to a water-borne silver metallic or red basecoat disintegrated after 4 years' exposure and could not be quantitatively analysed (see Table 4.24). In the case of the stabilized coatings, HALS-2 brought a slight improvement of gloss retention to the silver metallic basecoat. The adhesion between the clear coat and the basecoat was good even without basecoat stabilization (Gt 0).

In the case of the red basecoat, the presence of more HALS-2 in the basecoat had a definitely positive effect, gloss retention improved by more than 50 units. Adhesion of the formulation containing no additional light stabilizer was totally inadequate (Gt 5) but excellent when the basecoat contained an additional 1 % of HALS-2 (Gt 0).

Table 4.26: Effect of additional basecoat stabilization on the properties of a TSA clear coat

No.	Stabilization	4000 h exposure A			4300 h exposure B		
		20° gloss	ΔE (exposed/ unexposed)	adhesion[1]	20° gloss	ΔE (exposed/ unexposed)	adhesion[1]
1	Clear coat: unstabilized basecoat: unstabilized	cracking after 3200 h			84	2.0	Gt 3
2	Clear coat: 2 % BTZ-7/1 % HALS-2 basecoat: unstabilized	55	1.3	Gt 0	80	1.1	Gt 0
3	Clear coat: 2 % BTZ-7/1 % HALS-2 basecoat: 1 % HALS-2	67	1.1	Gt 0	81	0.7	Gt 0
4	Clear coat: 2 % BTZ-7/1 % HALS-2 basecoat: 1.5 % BTZ-7	60	1.2	Gt 0	84	0.5	Gt 0
5	Clear coat: 2.8 % BTZ-7[2)]/1 % HALS-2 basecoat: unstabilized	61	1.1	Gt 0	83	0.3	Gt 0

Basecoat water-borne, red, bake: 130 °C, 30'

1 according to DIN 53151 after additional exposure F; tape test to check intercoat adhesion clear coat/basecoat
2 the amount of UV absorber in the basecoat in formulation no. 4 was additionally incorporated in the clear coat

The effect on the dry film thickness on the light sensitivity of the underlying layers is shown in Table 4.25. A blue basecoat was applied with a film thickness of 10 µm, rather than the usual 20 µm, on a very light sensitive electrophoretic layer. Subsequently, a clear coat was applied with a dry film thickness of 30 µm, rather than 40 µm, to give a triple layer coating. After only 2 years of weathering in Florida: the unstabilized coating as well as that where only the basecoat contains UV absorber, complete delamination of the electrophoretic layer occurs. Such unstabilized clear coats are unable to protect the electrophoretic coating from UV light, despite the presence of light absorbing pigment and the UV absorber in the basecoat. This leads to degradation of the electrophoretic coating its delamination.

If the clear coat is stabilized with UV absorber and HALS but the basecoat unstabilized, adhesion is much better (Gt 2) since the harmful UV light is mostly blocked and only a very small amount reaches the electrophoretic layer. If UV absorber is added also to the basecoat whereby the remaining UV light is captured, adhesion is unchanged (Gt 0).

The same effect can be achieved by adding the total amount of the UV absorber directly to the clear coat (Table 4.26).

Analysis of the weathering A tests reveals that gloss retention is greatly improved when HALS-2 is added to the basecoat (cf. formulations No. 3 and 2). The addition of UV absorber to the basecoat or an increase in UV absorber concentration in the clear coat has no effect on coating stability (cf. Nos. 4 and 5 with No. 2). Furthermore, additional stabilizer in the basecoat does not improve gloss retention or adhesion.

Under weathering conditions B, additional stabilizer in the basecoat brings no improvement in gloss or adhesion (cf. Nos. 3, 4 and 5 with No.2). Differences in colour retention are, however, apparent: better colour retention is achieved through the use of HALS-2 and BTZ-7 in the basecoat (cf. Nos. 3, 4 and 5 with No.2). The effect obtained with BTZ-7 in the base coat is also can also be reproduced by a similar increase in the amount of UV absorber in the clear coat (cf. Nos. 5 and 4).

Therefore, if a car factory receives basecoat and clear coat from the same supplier, it is advisable to place all the UV absorber in the clear coat, where the screening effect is greatest. Additional stabilization of the basecoat with HALS may be advantageous, depending on the binder and pigment. If basecoat and clear coat are supplied by different manufacturers, the risk of insufficient stabilization, UV absorber and HALS, in the clear coat can be averted by adding extra light stabilizer to the basecoat. Since pigments also absorb UV light to a certain extent (see Chapter 3.1), tests should be carried out to ascertain the amount of additional stabilizer to be added to the basecoat. If the addition of more HALS is ineffective, the addition of UV absorber can be evaluated.

4.2.4.9 Exposure results for one-coat finishes

The principles of light stabilization of automotive finishes were discussed in Chapter 4.2. In particular, Chapter 4.2.1 focused on the underlying principles of stabilizing two-coat finishes. These principles must be modified for 1-coat finishes due to the presence of pigments. Figure 4.4 in Chapter 4.2.1 shows the fate of light when impinging on a clear coat applied on a basecoat. The former usually has a coating thickness of about 40 μm. Here, the incident light has many possibilities for reflection, as Figure 4.4 clearly shows, and a homogeneously dispersed UV absorber can filter out a large proportion of harmful UV light before this can reach the pigmented basecoat.

The conditions reigning in 1-coat finishes is very different. The pigment particles near the surface of the paint film are only covered by a very thin binder film and interact directly with the impinging UV light. As pigments can absorb some UV light (see Chapter 3.1), they augment the effect of true UV absorbers such as benzotriazoles, hydroxyphenyl-s-triazines, etc. However, the pigment is present in far higher concentrations than the UV absorber diminishing the latter's screening and stabilizing effect.

Of course, the stabilizing effect could be increased by simply adding more UV absorber. In Europe, it is normal to add 1.5 to 2 % UV absorber to a 40 μm thick clear coat. According to Lambert-Beer's law, the stabilizing effect or extinction E is equal to $\varepsilon \cdot c \cdot d$ (see Chapter 3.2.1). If E is to remain constant, a reduction in the coating thickness d must be accompanied by a similar increase in the UV absorber concentration c. The extinction coefficient ε

is a constant specific to the UV absorber. Considering a 1-coat finish where the unpigmented binder film on the very top of the coating is assumed to be 2 µm thick, the UV absorber concentration would have to be increased twenty times (!) in order to achieve the same stabilizing effect of the 40 µm clear coat. This results in a UV absorber concentration of 30 to 40 %, so high that it will seriously influence all other coating properties, for example the surface hardness or the mechanical properties.

Therefore, a *general* use of UV absorbers makes little economic sense for one-coat finishes. Unless, of course, it is warranted for special pigments or when the pigment/binder ratio is small, as in case of low pigment-volume concentrations (PVC).

The tell-tale sign of damage in 1-coat finishes is not the appearance of surface cracks but a general dullness of the surface caused by degradation of the binder. In extreme cases, chalking occurs since the binder has decomposed to such an extent that the pigment particles near the surface are no longer covered.

Figure 4.53 shows the colour change ΔE after weathering of a red, alkyd/melamine one-coat finish containing 2 % BTZ-8. The colour change is a good measure of the UV absorber's effectiveness. At the high pigment/binder ratio of 1 : 4, where very little pigment-free binder film is present, the UV absorber's screening effect is weak. The use of more binder (ratio of 1 : 8) leads to a thicker pigment-free binder film and better protection as seen by the 3 point improvement in colour change. Unlike UV absorbers, that efficacy of HALS, as far as gloss retention and colour retention are concerned, is not affected

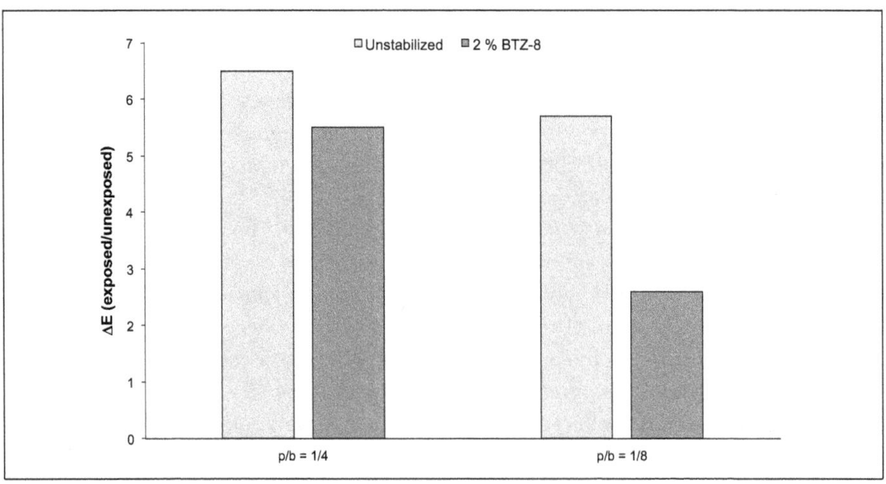

Figure 4.53: Effect of pigment/binder ratio (p/b) on UV absorber effect
binder: alkyd/melamine, pigment: Pigment Red 254 / TiO_2 50/50, bake: 130 °C, 30', exposure: 2 years Florida (5° S, bb, unheated)

Table 4.27: Stabilizing effect of HALS in a 1-coat alkyd/melamine coating

Pigment/binder ratio	**1/4**	**1/6**	**1/8**	**1/4**	**1/6**	**1/8**
Stabilization	20° gloss			ΔE (exposed/unexposed)		
Unstabilized	2	4	4	6.5	6.2	5.7
2 % HALS-3	40	53	59	1.9	1.2	0.8

pigment: Pigment Red 254/TiO_2 (50/50)
bake: 130 °C, 30'
exposure: 2 years Florida (5° S, bb, unheated)

by film thickness and is, thus, largely independent of the pigment/binder ratio (see Table 4.27). However, due to the presence of large amounts of pigments and other additives such as dispersing agents or potential adsorption of HALS on the pigment surfaces throughout the pigmented layer, it is advisable to use a slightly increased HALS concentration, for example 1.5 to 2 %.

Figure 4.54 demonstrates that the stability of a 2P-PU one-coat refinish is improved by an increase in HALS concentration from 0.5 % to 1.5 %. After 2 years' exposure in Florida, the unstabilized sample had become matt due to binder degradation. Gloss values after 3 years indicated that the pigment/binder system is more resistant to degradation at a HALS-2 concentration of 1.5 % than at 0.5 %.

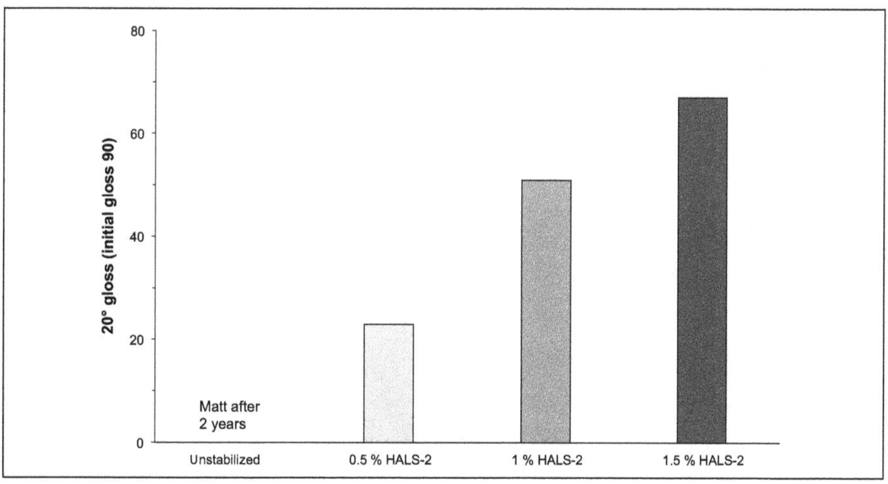

Figure 4.54: Effect of HALS on gloss retention in a blue 1-coat 2P-PU coating
binder: acrylic polyol/isocyanate bake: 80 °C, 45'

The effect of HALS on pigmented coating stability depends not only on the binder, but also on the pigment. If pigments are not weather resistant, the stabilization of the binder/pigment system in one-coat finishes will be difficult, even in the presence of HALS and/or UV absorbers (see Figure 4.55).

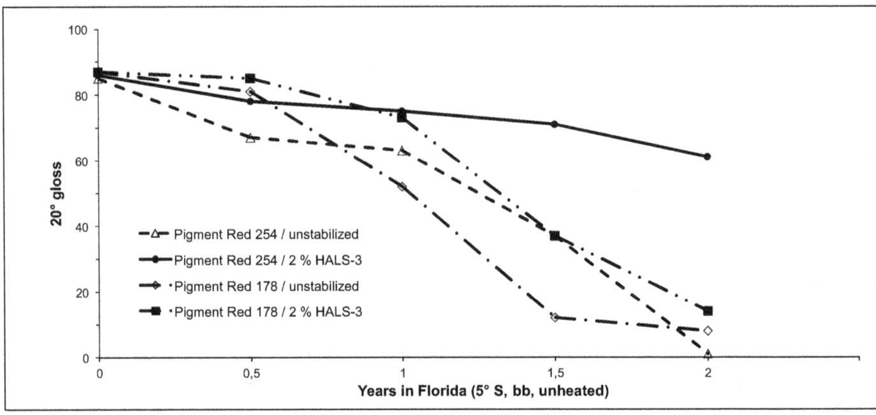

Figure 4.55: Outdoor exposure of a red 1-coat alkyd/melamine coating, p/b = 1/6
bake: 130 °C, 30'

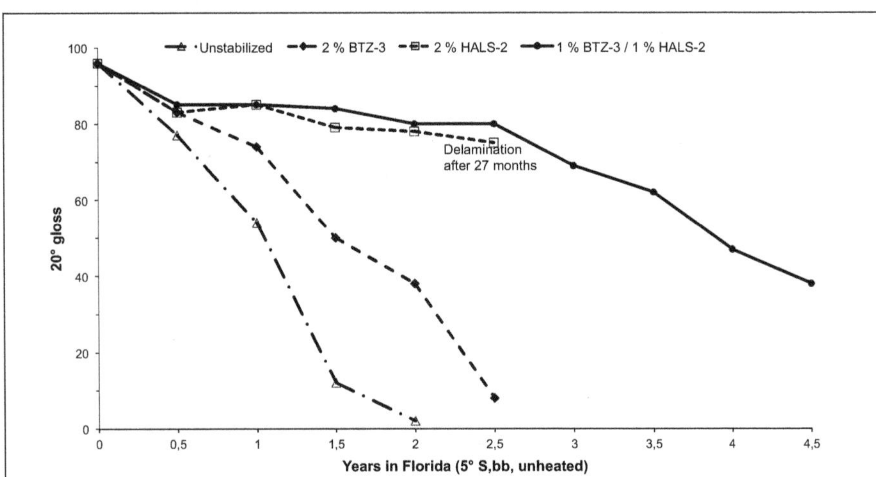

Figure 4.56: Effect of UV absorber and HALS on weathering resistance of a 1-coat 2P-PU coating
binder: acrylic polyol/isocyanate, pigment: Pigment Blue 28, bake: 80 °C, 45', substrate: epoxy primer

Two different red pigmented alkyd/ melamine coatings (p/b 1:6), unstabilized and stabilized with 2 % HALS-3, were subjected to outdoor weathering. After 2 years of weathering in Florida, the unstabilized coating containing Pigment Red 178 proved to have the worst weathering resistance. A slight improvement was achieved by adding 2 % HALS-3 but this coating was no better than the unstabilized coating containing Pigment Red 254, a more stable pigment. The best result by far was achieved with the coating based on Pigment Red 254 stabilized with 2 % HALS-3. These results clearly show that the weather resistance of the pigment itself determines to a great extent what level of stabilization is possible.

As mentioned earlier, the use of a UV absorber in a one-coat finish can make financial sense depending on the pigment/binder ratio (Figure 4.56).

In contrast to the analogous system shown in Figure 4.54, the pigmented 2P-PU refinishing was based on the blue pigment cobalt spinel, which is transparent to UV light. Upon exposure in Florida for two years, the sample without light stabilizer had become matt (similar to the system in Figure 4.54). As expected for UV absorbers in one-coat systems, use of 2 % BTZ-3 brought only a slight improvement in gloss retention and a minimal increase in stability: the surface had become matt after 2½ years. On the other hand, 2 % HALS-2 led to high gloss retention over 2½ years but did not prevent delamination. This results from the transparency of the cobalt spinel pigment, which allows almost free passage of the UV light to the underlying epoxy primer. The latter is UV sensitive and degrades leading ultimately to loss of adhesion. Best results were obtained with a blend of 1 % HALS-2 and 1 % BTZ-3. HALS protects the polyurethane binder against degradation and ensures very good gloss retention. BTZ-3 does not have any appreciable effect on gloss retention but, by filtering out harmful UV light, protects the epoxy primer from degradation and preserves adhesion.

The following conclusions can be made concerning the stabilization of one-coat finishes: HALS are indispensable since they protect the binder from gloss deterioration and chalking and should be added in amounts between 1.5 and 2 %, based on solid binder. Depending upon the type of pigment and the pigment content, the use of additional stabilizer, for example a UV absorber, *may* be beneficial.

4.3 Light stabilization of industrial coatings

This Chapter focuses on coatings used in industrial applications other than automotive. In automotive finishing, several coating layers are applied (phosphating, electrophoretic paint, filler, basecoat, clear coat) to protect the metal from damage and corrosion, as well as to create a decorative effect. The prime function of industrial paints is the protection of the substrate.

Industrial paints are applied to rolling stock, aircraft, ships and boats, agricultural vehicles and machinery, construction, wooden components and many other objects. UV absorbers and HALS are also essential in extending the life of such coatings [190, 191].

The most important difference between industrial and automotive finishes lies in the curing conditions. Curing temperatures for automotive finishes are typically between 120 and 140 °C, whereas industrial finishes can be cured at anything between 20 °C, for example two-pack paints or oxidative curing coatings, and 300 °C, coil coatings for example. Today, suitable light stabilizers are available for all these applications and temperature ranges.

The basic principles and criteria for selecting the right light stabilizers for industrial paints are similar to those for automotive finishes. Therefore, the know-how for the stabilization of automotive finishes can also largely be applied to industrial paints. Several examples will be discussed in this chapter, with a focus on the protection of metal and wood substrates.

4.3.1 Stabilization of paints for metal substrates

An unstabilized ambient cured red two-pack epoxy coating starts to show signs of chalking (degradation of the binder, which exposes the pigment particles) after 2200 hours of weathering under conditions E (Figure 4.57). After 2400 hours, chalking is very pronounced.

Significant improvement in the weathering resistance comes with the addition of 1 % HALS-3; no signs of chalking are apparent even after 2800 hours of weathering. Experi-

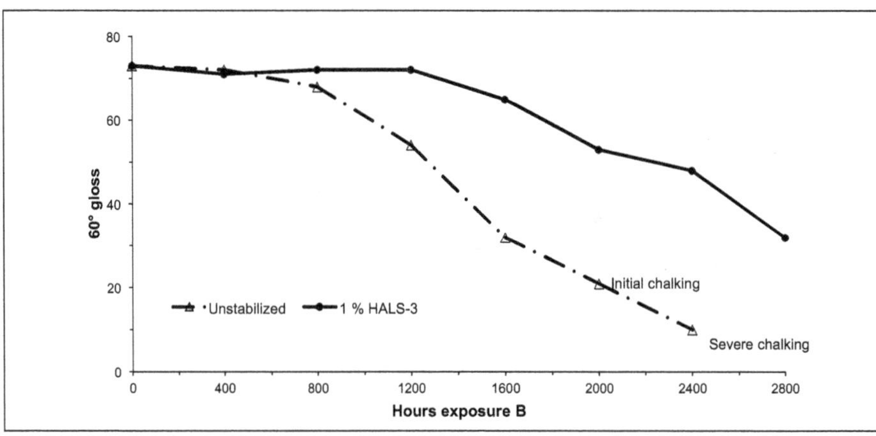

Figure 4.57: Accelerated weathering of an air-dried red pigmented 1-coat 2P-epoxy coating

ence with one-coat automotive finishes suggests that enhanced stabilization would result from a loading of 1.5-2 % HALS-3.

Water penetration leading to corrosion of the underlying substrate can occur when capillaries form in the damaged binder network. Protection of the binder with HALS and/or UV absorber can therefore delay or prevent water penetration into the paint film, thus improving corrosion resistance. This is exemplified in Figure 4.58 where a white, ambient cured two-pack PU paint was first exposed for 21 months in Florida (45° S) then to salt spray for 525 hours' exposure (DIN 50021 SS). As expected, the paint stabilized with only 1 % HALS-2 shows better corrosion resistance than the unstabilized paint.

Figures 4.59 and 4.60 show the weathering of a blue pigmented coil coating system prepared by applying a thermoplastic acrylate/polyvinylidene fluoride binder to an epoxy primer and curing for 1 minute at 260 °C. Whereas the unstabilized sample showed pronounced signs of delamination after 3½ years exposure in Florida, that stabilized with a BTZ-7/HALS-2 blend was not only intact after 4 years weathering but showed excellent gloss (85° gloss = 65 units, Figure 4.59).

The loss of adhesion in the unstabilized sample is due to the photo-chemical degradation of the epoxy binder in the primer; the blue pigment (cobalt spinel) in the top coat is too transparent to filter out the UV light. The only solution is to use a blend of UV absorber and HALS, as discussed in Chapter 4.2.4.9 (Figure 4.56). The appearance of the stabilized paint formulation after 51 months of outdoor weathering is illustrated in Figure 4.60.

Industrial powder coatings can be used for both exterior and interior applications. Until the second half of the nineties, coatings based on polyester/TGIC (triglycidyl isocyanurate) were considered the state of the art for outside use [192]. As a result of the

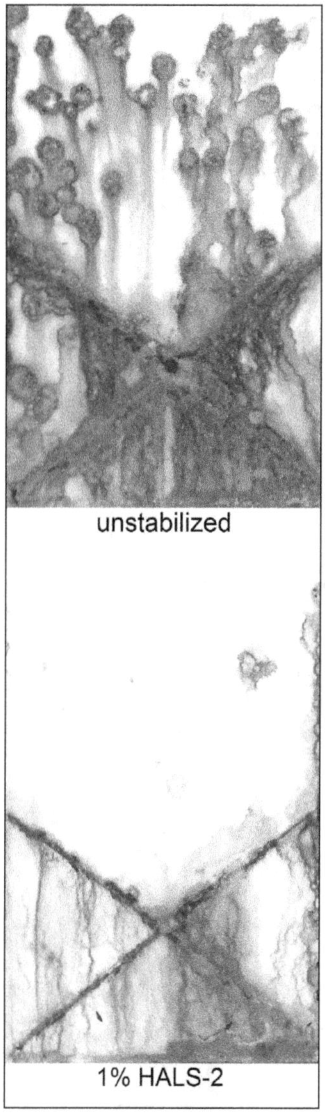

Figure 4.58: Effect of HALS on the corrosion resistance of a 1-coat system
binder: acrylic polyol/isocyante, pigment: TiO_2, exposure: 21 months Florida (45° S) + 525 h salt spray test (DIN 50021 SS)

reclassification of this hardener, TGIC-free powder coatings for exterior applications are gaining ground, for example by using alternative hardeners such as hydoxy-alkylamides. Nevertheless, TGIC based systems are still being used for both general industrial applica-

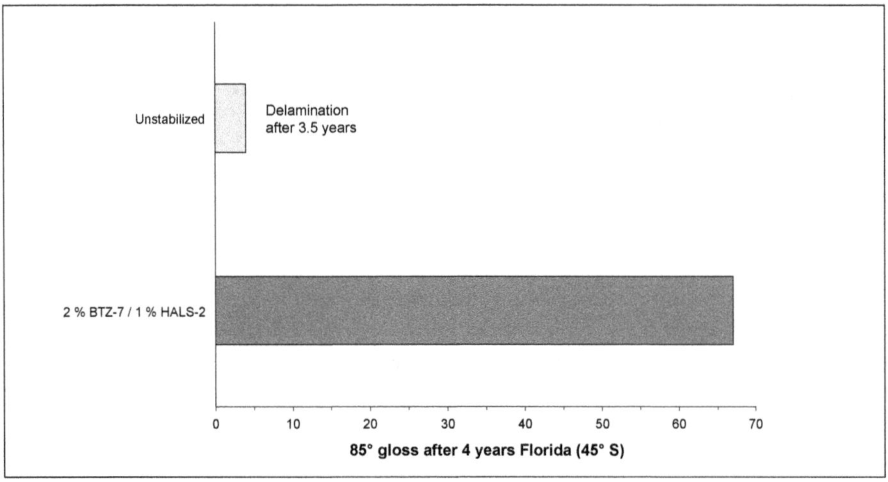

Figure 4.59: Outdoor exposure of a 1-coat coil coating
binder: thermoplastic acrylate/polyvinylidene fluoride, DFT: 30 µm, pigment: Pigment Blue 28, bake: 260 °C, 1', substrate: epoxy primer, DFT: 6 µm

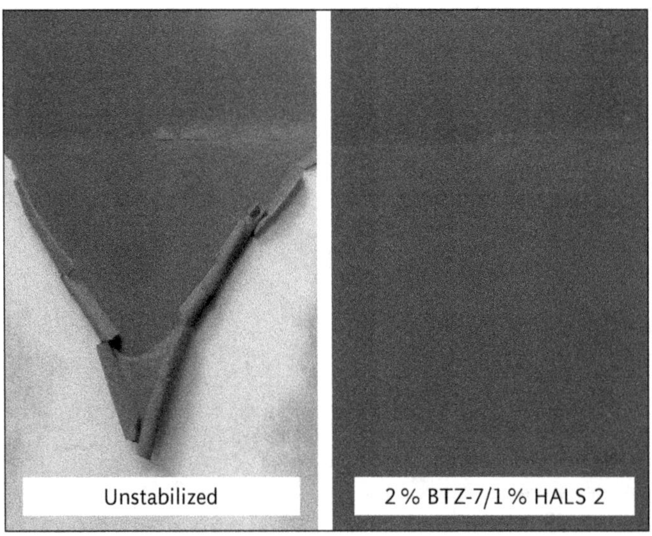

Figure 4.60: 51 months' outdoor exposure of a 1-coat coil coating
binder: thermo-plastic acrylate/polyvinylidene fluoride, DFT: 30 µm, pigment: Pigment Blue 28, bake: 260 °C, 1', substrate: epoxy primer, DFT: 6 µm, exposure: Florida (5 °S, bb, unheated)

tions and for so called super-durable systems. In general, the lightfastness of such systems can be improved by adding light stabilizers [193]. Figure 4.61 shows the improvement in stability of a pigmented white polyester/TGIC powder one-coat system possible through use of UV absorbers and HALS.

The gloss retention of the unstabilized paint has decreased considerably after 1600 hours of weathering (conditions B) and first signs of chalking are evident after 2800 hours. After 3200 hours, very pronounced chalking is present. If the paint is stabilized with only 1.5 % HPT-2 or 1.5 % HALS-4, gloss retention and chalking are similar to the unstabilized sample. Only by combining HPT-2 with HALS-4 is it possible to improve gloss retention and prevent chalking.

4.3.2 Stabilization of clear wood coatings

Despite increased use of materials such as plastics and the like, wood remains a popular substrate for both decorative and construction purposes. Common applications include furniture, joinery, parquetry, timber work and sun decks.

Unlike metal, which is a "dead" material, wood is living. To ensure durability and long life it is typically coated with decorative and protective finishes such as paints, transparent stains, penetrating finishes or film forming coatings [194]. In outside use, it is not only exposed to photo-chemical and physical influences but is also likely to be attacked by mould and insects. All these factors must be taken into account in the choice of the paint or protective

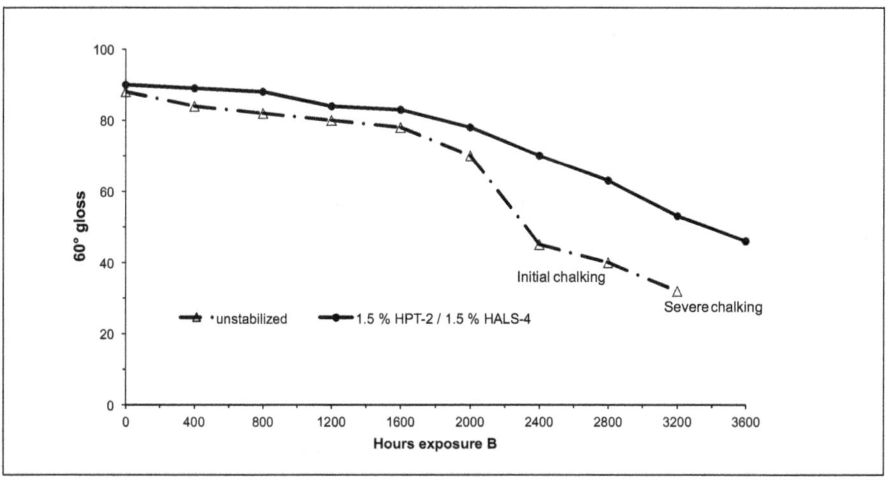

Figure 4.61: Accelerated weathering of a white pigmented, 1-coat powder coating
binder: polyester/TGIC (93/7); DFT: 70 µm, bake: 195 °C, 15'

coating[195, 196]. Until now, mainly pigmented paints were used to protect wood in exterior applications[197]. Of these, iron oxide pigments found particularly wide use[198, 199]. Since the mid-eighties, there has been a shift towards transparent systems, which reveal the natural features of wood such as its colour, grain and texture. However, due to the UV light transparency of the topcoats and the extreme light sensitivity of certain wood components such as lignin, the stability of transparent systems is rather limited in exterior applications. Even under diffuse interior light conditions, pale wood tends to exhibit yellowing and darkening as a result of the photo-oxidation of lignin[194]. Such coatings can only be used if they are adequately stabilized with UV absorbers and HALS[200–203]. Further improvement can be achieved by pretreating the wood with a specific lignin stabilizer such as HALS-12. The latter will be discussed after the following examples of conventional stabilization of fir, pine and beech with UV absorber and HALS.

Figure 4.62 shows the results in a solvent based long-oil alkyd (LOA) clear coat system on fir after 18 months' exterior exposure (45° North in Sydney, Australia).

The superior performance of HPT-6 vs. BTZ-8 and that of HALS-3 vs. HALS-2 is clearly evident. In case of the UV absorbers, the higher photo permanence and the broader spectral coverage of HPT-6 brings greater stabilization. The basicity of the HALS explains the superiority of HALS-3: the more basic HALS-2 tends to interfere with the oxidative curing process leading to cure retardation and subsequently inferior performance.

A similar system on pine is summarized in Table 4.28, where colour change ΔE and yellowness index change ΔYI after 2 years' exterior exposure (45° South) near Basle, Switzerland were measured.

The unstabilized reference sample shows a distinct colour change, as reflected in the yellowness index. Addition of 2.5 % HALS-3 to the clear coats provides slightly better results in both cases, whilst benzotriazole-5 further improves colour retention. However, the

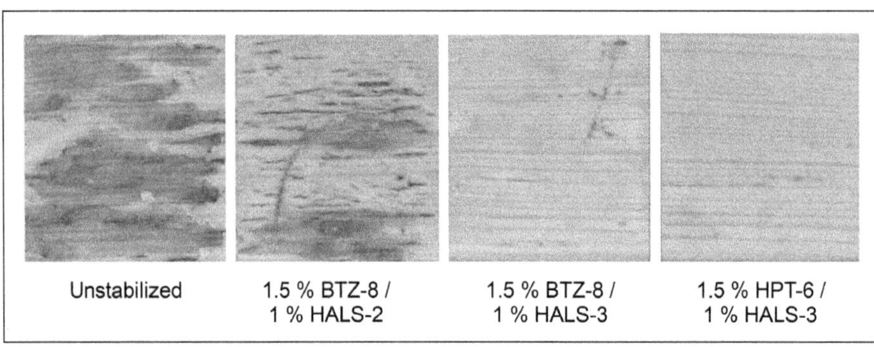

Figure 4.62: Exterior exposure of an air dried clear wood coating, clear coat: long-oil alkyd, solvent based, substrate: fir, exposure: 18 months Sydney, Australia (45° North)

Table 4.28: Outdoor exposure of a long-oil alkyd based clear coat

Stabilization	**ΔE (exposed/unexposed)**	**ΔYI (exposed/unexposed)**
Unstabilized	15.2	28.7
2.5 % HALS-3	8.0	15.0
2.5 % BTZ-8	4.6	8.9
1.5 % BTZ-8/1 % HALS-3	1.0	2.6

substrate: pine, exposure: 2 years Basle (45°S)   colour change ΔE; yellowness index ΔYI

Table 4.29: Wood with and without lignin stabilization

Stabilization	**Δ E after .... hours**		
	240 h	**960 h**	**2160 h**
Unstabilized (with/without pretreatment)	30/30	cracking	
BTZ-7 (no pretreatment)	11	22	cracking
BTZ-7 (**with** pretreatment)	4	10	19
LS-Blend 2 (no pretreatment)	10	20	26
LS-Blend 2 (**with** pretreatment)	3	5	20
LS-Blend 3 (no pretreatment)	8	13	33
LS-Blend 3 (**with** pretreatment)	4	4	8

clear coat: water-borne acrylic dispersion   pretreatment: 2 % aqueous solution of HALS-12   exposure: exposure G

best results are obtained with a combination of 1.5 % benzotriazole-8 and 1 % HALS-3 where the changes in colour, ΔE (1 unit) and yellowness index ΔYI, (2.6 units) were the smallest.

Wood, painted or unpainted, is also subjected to cyclic accelerated weathering (see Table 4.17, Exposure G)

As aforementioned, a significant increase in the stabilization of pale wood can be achieved through its pretreatment with a specific lignin stabilizer such as HALS-12. Table 4.29 shows the improvement in colour change ΔE attainable through pretreatment of pine with a 2 % aqueous solution of HALS-12 before application of a water-based acrylic clear coat stabilized with UV absorber/HALS blends (see Table 4.11).

The data clearly show that additional lignin stabilization via pretreatment of the wood and the appropriate UV absorber/ HALS combination in the coating results in superior colour retention of the substrate. It is also important to note that fungicides used in wood coatings are also protected by light stabilizers to prolong their longevity [204].

For highly pigmented (white or coloured pigments) top coats for wood, the comments made in Chapter 4.2.4.9 apply. Since the pigment also acts as UV absorber, the system should be stabilized with 1.5 to 2 % HALS, based on solid binder.

UV absorbers can also be used to prevent wood discoloration in interior applications. Here, 1.5 to 2 % of UV absorber should be added to the clear coat, based on solid binder. Although HALS is not usually necessary in such interior applications, they can prevent yellowing of the binder in pigmented finishes and, in the case of delicate binders, prevent chalking.

For UV-cured coatings for furniture and parquet flooring, the principles discussed in Chapters 4.2.2.6 and 4.2.4.5 apply.

4.4 Stability of light stabilizers

Light stabilizers must be stable so as to provide protection over a long period of time. Three factors determine the stability of a light stabilizer: **volatility** at elevated temperatures, **photo-chemical stability** and **extraction resistance** [44, 205, 206]. Volatility and extraction resistance have already been discussed in Chapters 4.2.2 (UV absorber) and 4.2.3 (HALS). Thus, the remaining factor is the *photo-chemical* stability of UV absorbers and HALS.

4.4.1 Photo-chemical stability of UV absorbers

The photo-chemical stability of UV absorbers has been the subject of much research [206–223]. *Gerlock et al.* [212] investigated the factors which contribute to the decomposition of UV absorbers using BTZ-3 as example. Although benzotriazole UV absorbers are resistant to direct photolysis [213], in the presence of oxygen, they can take part in photo-oxidation reactions, depending on the type of binder used. In the ground state, the benzotriazole ring system is planar due to the intra-molecular hydrogen bridge and does not react with radicals. According to *Gerlock et al*, the twisted excited state is susceptible to radical attack and leads eventually to the decomposition of the benzotriazole molecule. *Decker* [209, 210] proposes two pathways for UV absorber decomposition: direct photolysis, which is contrary to the findings of *Catalan et al.* [213], and radically induced photo-oxidation.

Not surprisingly, the long-term stability of UV absorbers in coatings has also been investigated. Two non-destructive methods have been used for the analysis:

- UV spectroscopy of clear coats: the formulation is applied to a quartz plate and thermally cured. The coatings are then subjected to artificial weathering. The change in the UV absorber content over time is derived from measurement of the extinction of the coating at the absorption maximum λ_{max} of the UV absorber. To minimize experimental error (due to variations in coating thickness), the extinction measurements are always made at the same spot.
- UV reflectance spectroscopy [214]: unlike method 1, this is a *semi-quantitative* method, which provides no information about the absolute UV absorber content. On the other hand, it is easy to perform and allows a good assessment of the UV stabilization. This method has been used for test panels subjected to outdoor exposure and is especially suitable for clear coat/basecoat systems in which the pigment particles have little or no absorption at 300 to 400 nm.

Figure 4.63 illustrates the stability of HPT-1 in two different acrylate/melamine-based clear coats applied to a quartz plate [215]. Here, as described earlier, spectroscopically examined at regular intervals. The weathering conditions comprised continuous irradiation in a UVA-340 at 40 °C in the absence of moisture and assessment was made according to the time resulting in a 50 % loss of UV absorber. The concentration of HPT-1 chosen gives an optical density of 1.53 at 340 nm.

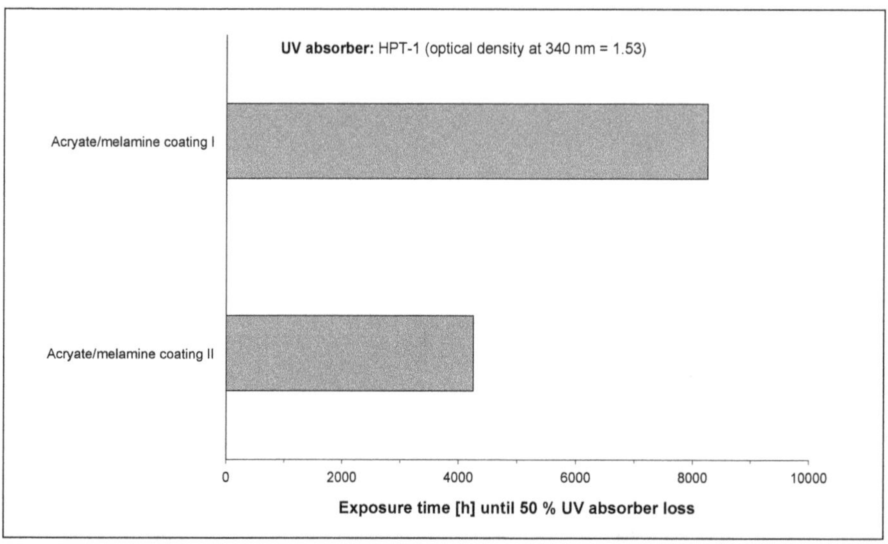

Figure 4.63: Effect of different acrylate /melamine binders on the stability of UV absorbers [215]
bake: 130 °C, 30', DFT: 35 μm exposure: UVA-340 bulbs; continuous irradiation at 40 °C (dry conditions)

It is evident that the stability of HPT-1 is lower in the acrylate/melamine paint II, even though both systems have the same acrylate/melamine ratio. These findings confirm those of *Gerlock et al.* [216], who concluded that the method used to synthesize the polyacrylate (for example, choice of solvent) has a strong influence on the generation of radicals upon aging of the cured paint film. The more radicals, the greater the probability of UV absorber decomposition through radical attack.

Figure 4.64 compares the effect of different crosslinking mechanisms and binders on UV absorber stability. The UV absorber used was BTZ-3. Clearly, the stability of BTZ-3 increases as the binder system is changed from high-solids acrylate/melamine to 1P- PU to fluoropolymer/melamine. Fluoropolymers are known for their high stability toward radical decomposition. Thus, fewer free radicals are formed and less radical induced decomposition of the UV absorber occurs. If radical generation within the polymer has a pronounced effect on the stability of UV absorbers, the two following tests should deliver the confirmation.

- A change in radiation source with the concomitant change in the radiative energy should result in a different radical yield in the paint film, leading to a difference in UV absorber stability.

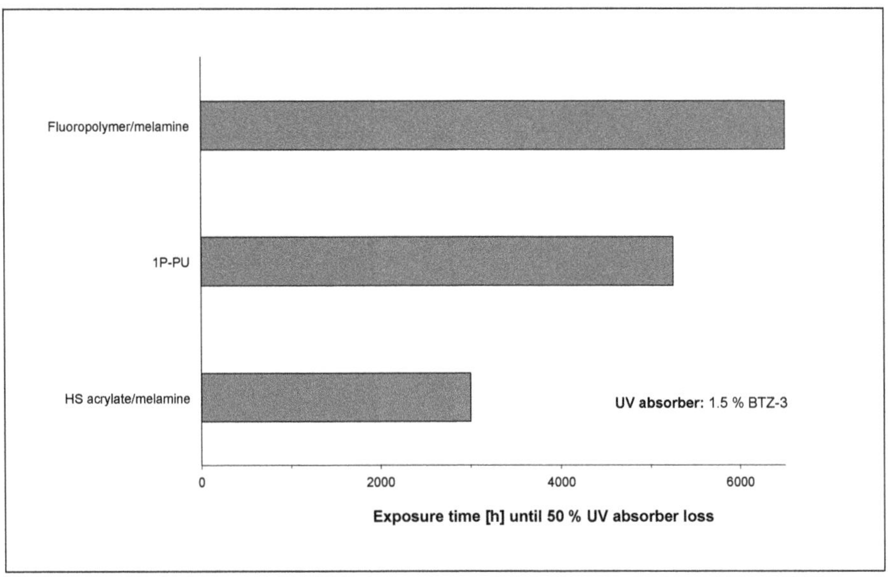

Figure 4.64: Effect of binder on the stability of UV absorbers [215]
DFT: 30 μm, exposure: UVA-340 bulbs; continuous irradiation at 40 °C (dry conditions)

- The addition of HALS, free-radical scavengers, should intercept the radicals before they can attack the UV absorber and also bring an increase in UV absorber stability.

Indeed, the data clearly show that the shorter and more energetic wavelengths emitted by the UVB-313 lamps relative to UVA-340 lamps lead to faster decomposition of BTZ-8 (Figure 4.65). It is a well-known fact that use of the UVB-313 lamps also leads to higher peroxide radical concentrations [83]. Furthermore, the addition of HALS-2 significantly increases the stability.

In addition to these factors, clear coat thickness has a considerable influence on UV absorber stability. This was confirmed by *Decker* [209, 210] in a study using BTZ-3. The film thickness is directly related to the screening effect of the UV absorber. Given a homogeneous distribution of the UV absorber throughout the coating, the UV absorber in the top layers of the film will absorb the UV light, thereby protecting the UV absorber below. The thicker the clear coat film, the more pronounced the greater the protection. This is easily seen in an acrylic/melamine coating containing BTZ-7 applied to a silver metallic basecoat (Figs. 4.66 and 4.67). In a 20 µm thick clear coat, the decomposition of BTZ-7 over 4000 h

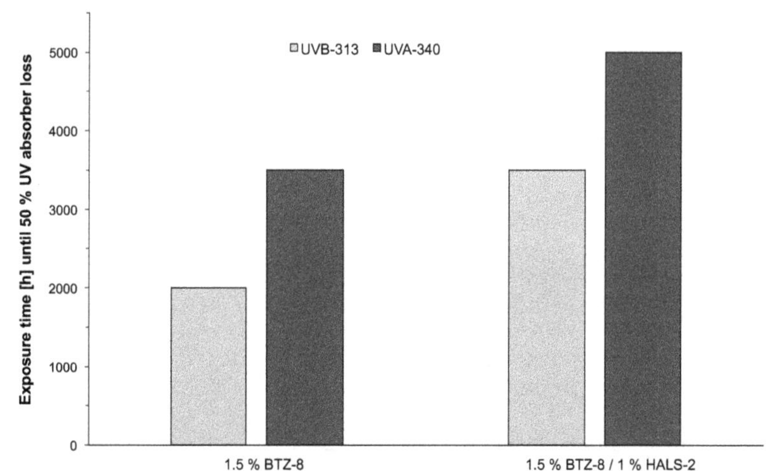

Figure 4.65: Effect of radiation type and HALS on UV absorber stability [215]
clear coat: 2P-PU (acrylic polyol/isocyanate), bake: 130 °C, 30', DFT: 35 µm, exposure: UVB-313 or UVA-340 bulbs; continuous irradiation at 40 °C (dry conditions)

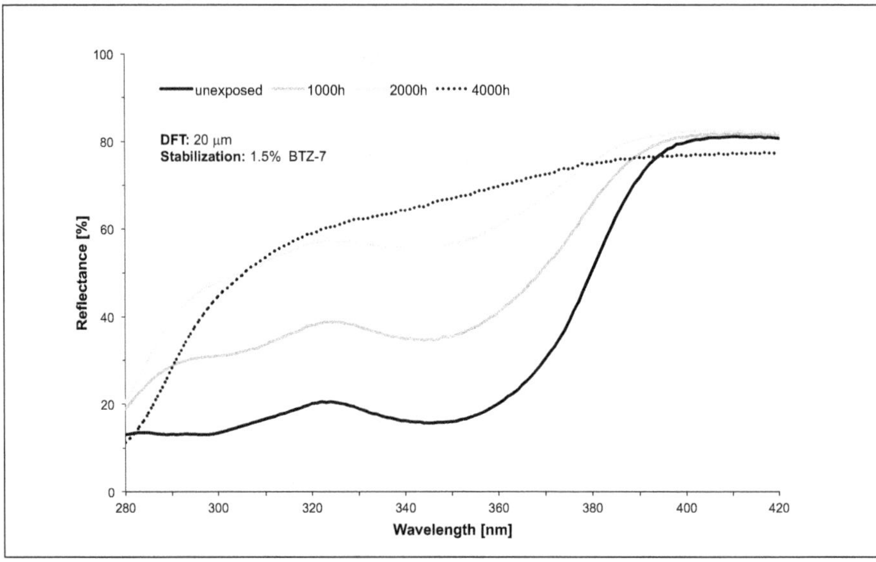

Figure 4.66: Reflectance spectra of a TSA clear coat
basecoat: silver metallic, bake: 130 °C, 30', exposure: exposure A

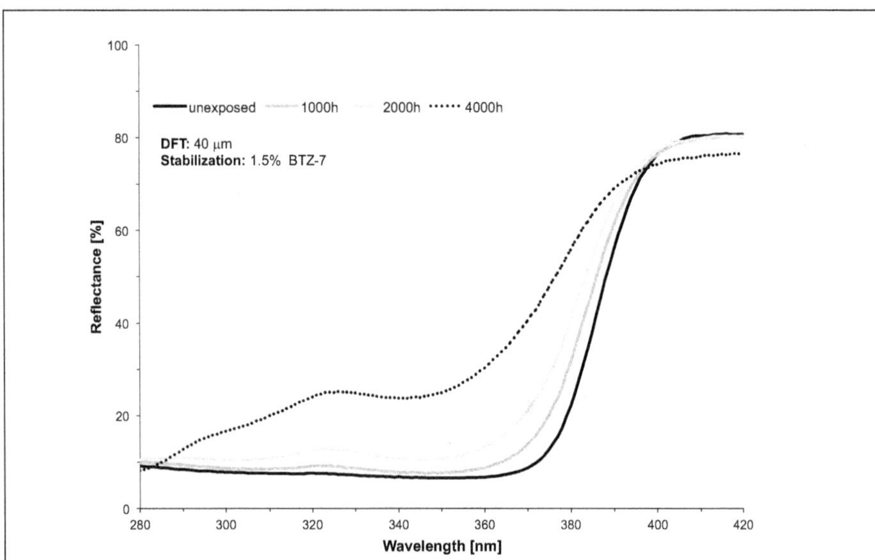

Figure 4.67: Reflectance spectra of a TSA clear coat
basecoat: silver metallic, bake: 130 °C, 30', exposure: exposure A

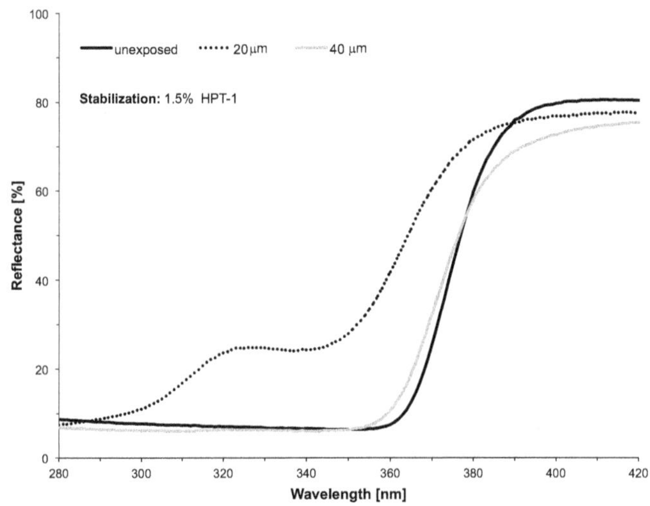

Figure 4.68: Reflectance spectra of a TSA clear coat after 4000 h exposure A
basecoat: silver metallic, bake: 130 °C, 30'

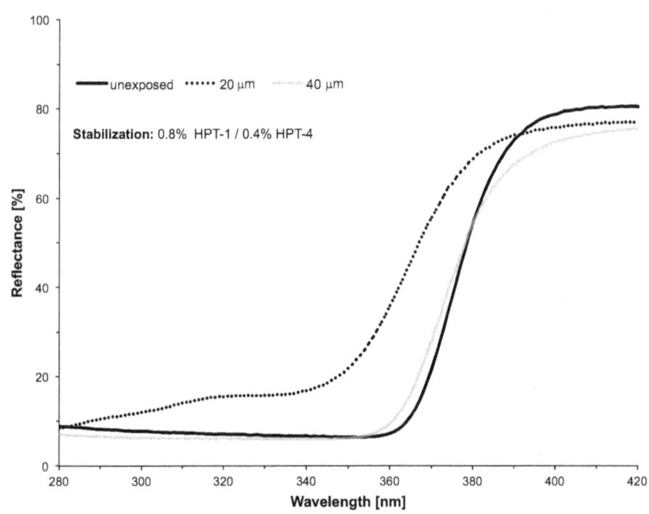

Figure 4.69: Reflectance spectra of a TSA clear coat after 4000 h exposure A
basecoat: silver metallic, bake: 130 °C, 30'

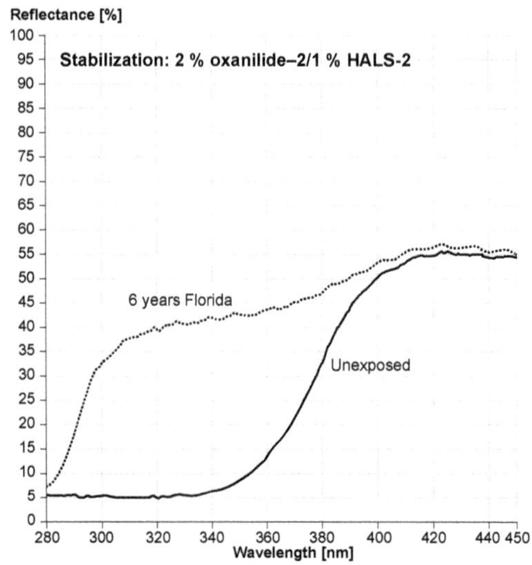

Figure 4.71: Reflectance spectra of a TSA clear coat
basecoat: light blue metallic, bake: 130 °C, 30', exposure: Florida (5° S, bb, unheated)

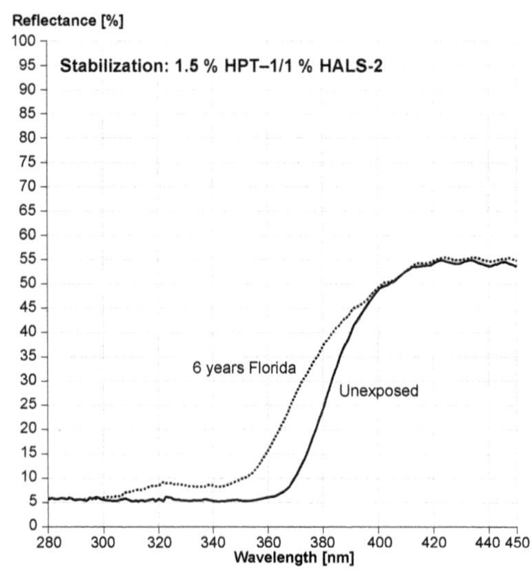

Figure 4.70: Reflectance spectra of a TSA clear coat
basecoat: light blue metallic, bake: 130 °C, 30', exposure: Florida (5° S, bb, unheated)

irradiation is significant. In contrast, some decomposition is observed in the 40 μm thick clear coat but it is significantly lower at each time interval compared to the 20 μm coating.

Clearly, the chemistry of the UV absorber molecule itself has a major influence on its photo-chemical stability (see Figures 4.68 and 4.69). Figure 4.68 shows the reflectance spectra of similar 20 μm and 40 μm thick clear coats containing HPT-1 after 4000 h of exposure (conditions A). Comparison of the 20 μm clear coat in Figure 4.68 after 4000 h of weathering with the corresponding curve in Figure 4.66 shows clearly that BTZ-7 has much lower stability than HPT-1: BTZ-7 has undergone complete decomposition. The better stability of HPT-1 is also evident from analysis of the 40 μm films but, at this larger film thickness, some BZT-7 remains, similar to the 20 μm clear coat with HPT-1.

Figure 4.69 shows the reflectance spectra of a combination of HPT-1 with HPT-4 in the same coating system. HPT-4 has prossess a significantly higher extinction coefficient and blue-shifted absorption maximum relative to HPT-1 (see Table 3.2 and Figure 3.13. in Chapter 3.2.3; Figure 3.13 shows the spectrum of HPT-2). This allows a reduction in the absolute UV absorber concentration relative to HPT-1 alone without loss in protection. In fact, comparison of the curves for the 20 μm thick clear coat of Figure 4.69 with the corresponding curve in Figure 4.68 reveals that an increase in photo-chemical stability is achieved through this combination.

The artificial weathering results are mirrored in the outdoor exposure tests. Figures 4.70 to 4.73) show the reflectance spectra of different groups of UV absorbers before and after weathering in Florida (USA) and Allunga (Australia).

According to the literature, benzophenones and oxanilides are less photo-chemically stable than benzotriazoles or hydroxyphenyl-s-triazines [99, 125, 153, 209]. A direct comparison of oxanilide-2 to HPT-1 in an acrylate/melamine clear coat (similar to coating II with increased radical yield in Figure 4.63) applied to a light blue metallic basecoat is shown in Figures 4.70 and 4.71.

Whereas the oxanilide-2 had completely decomposed after 6 years Florida exposure, HPT-1 still provided excellent UV stabilization. This is also reflected in the colour retention values: the system stabilized with oxanilide-2 showed a colour change ΔE of 4.4 units, whereas only a minor colour change ΔE of 0.6 units was observed for the HPT-1 stabilized film. Figures 4.72 and 4.73 depict the weathering behaviour of HPT-1 and BTZ-3 in a 2P-PU coating after 5.5 years' exposure in Allunga.

The spectra show that the photo-chemical stability of HPT-1 (Figure 4.72) is significantly higher compared to BTZ-3 (Figure 4.73). Again, the better filter effect leads to better protection of the underlying layers. The clear coat/basecoat adhesion, determined through the tape test after exposure to humidity for 96 h (exposure F), was excellent in the case of HPT-1 (Gt 0). This indicates that sufficient UV absorber remains in the lower layers of the clear coat. In contrast, the clear coats containing BTZ-3 suffered delamination (Gt 5) due to depletion of this UV absorber by photo-chemical degradation.

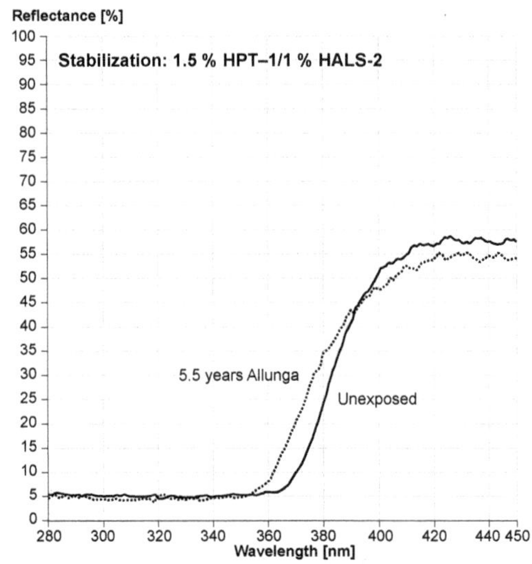

Figure 4.72: Reflectance spectra of a 2P-PU clear coat
basecoat: light blue metallic, bake: 80 °C, 30', exposure: Allunga (5°N, bb, unheated)

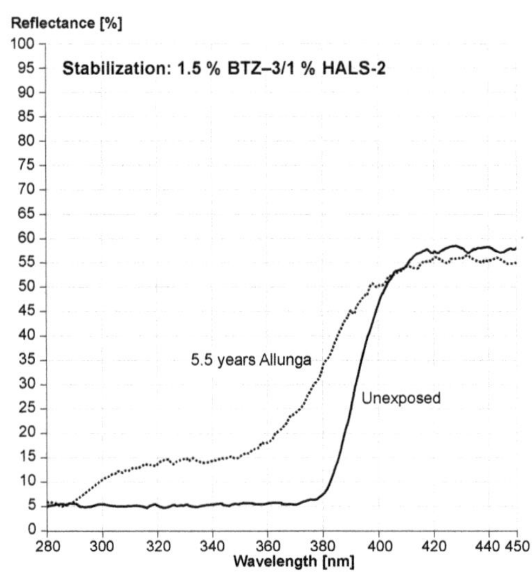

Figure 4.73: Reflectance spectra of a 2P-PU clear coat
basecoat: light blue metallic, bake: 80 °C, 30', exposure: Allunga (5°N, bb, unheated)

These results are in harmony with the technical literature, where the stability of the four major UV absorber groups decreases in the order of hydroxyphenyl-s-triazines ≥ hydroxyphenylbenzotriazoles >>> oxanilides ≈ hydroxybenzophenones.

4.4.2 Long-term stability of HALS

Analytical investigations of the long-term stability of HALS are challenging (see Figure 3.20 in Chapter 3.3.2.1). The only materials in equations (a) to (c) in Figure 3.20 which are *unequivocally* and easily characterized are >N-R (equation (a)) and the nitroxyl radical >NO·. The first, >N-R (provided it has not been built into the molecule, for example by copolymerization) can be quantified by extraction and analysis, the second, >NO·, by electron spin resonance (ESR) spectroscopy [62]. All the other species/products postulated in equations (a) to (c) are not precisely defined and virtually impossible to quantify in systems comprising binder, crosslinking agent, additives, pigments and fillers.

During the lifetime of HALS products, the >N-R moiety is fully converted into >NO· or other (intermediate) products. The reactions represented by equations (b) and (c) in Figure 3.20 are cyclic. Thus, the nitroxyl radical concentration determined at a certain point represents only a part of the total HALS concentration at that moment. *Gerlock* and *Bauer et al.* have devoted much effort to the analytical determination of nitroxyl radicals with the goal of predicting the long-term performance of HALS products. They found that the consumption of HALS is governed by the paint formulation (acrylate/melamine or 2P-PU, for example), the intensity of the radiation source and humidity [217]. These parameters are also relevant to UV absorber stability. The lifetime of HALS is longer in 2P-PU coatings than in those based on acrylate/melamine resins [218].

Experiments were performed under different accelerated weathering conditions and weathering resistance of up to 10,000 hours was observed [86, 218]. In all cases, the conclusions made were based on the >NO· concentrations measured. Even though artificial weathering trials indicate that HALS and their >NO· derivatives have long lifetimes, this still has to be confirmed by analytical data from outdoor weathering experiments. Figure 4.74 shows the nitroxyl radical concentrations expressed in 10^{-10} mol · spin per mg of paint when using 1 % HALS-2 in a 2P-PU clear coat.

The weathering tests were carried out under exposure A and C conditions. In the weathering tests carried out under conditions A, nitroxyl radical concentrations were more or less constant between 250 and 1000 hours weathering. This implies a cyclic stabilizing mechanism and that more >NO· is formed than is needed to intercept radicals under these conditions. Under weathering conditions C (short-wave, energy-rich), more >NO· is needed to intercept radicals than can be formed. This is reflected by the decrease in the nitroxyl radical concentration during the 250 to 1000 hours weathering period. The same clear coat containing an additional amount of BTZ-7 on a silver metallic base paint serves as a

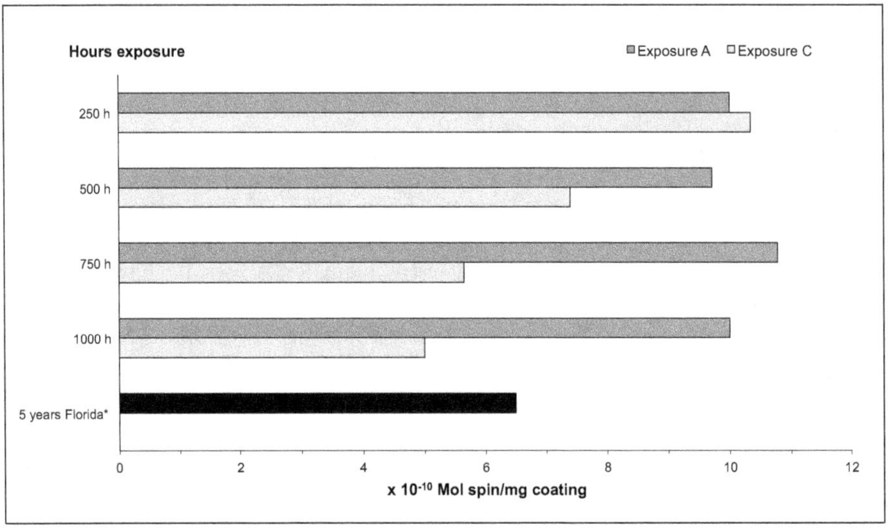

Figure 4.74: Long-term effect of HALS-2 in a 2P-PU
clear coat: acrylic polyol/isocyanate, DFT: 40 µm

* on top of a silver metallic basecoat; clear coat contains 2 % BTZ-7; exposure 5° S, bb, unheated

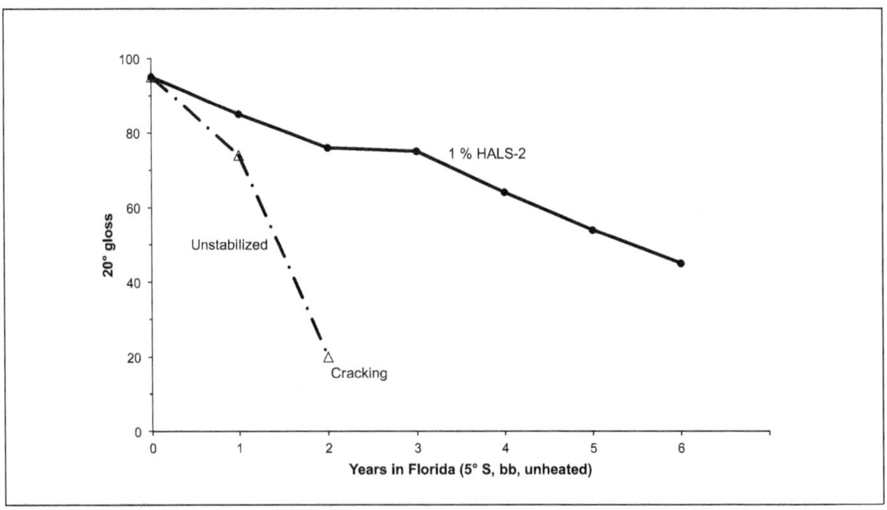

Figure 4.75: Long-term stabilization with HALS-2 in a 2P-PU
clear coat: acrylic polyol/isocyanate, basecoat: silver metallic, bake: 80 °C, 45'

comparison. After 5 years of weathering in Florida, enough nitroxyl radicals are still available to guarantee good stabilization. This concentration is higher than that obtained after 750 and 1000 hours weathering under conditions C.

The good long-term stabilizing effect of HALS-2 in a clear coat is illustrated in Figure 4.75. As pointed out earlier, HALS are, above all, effective for good gloss retention and crack prevention.

After 6 years outdoor weathering, no cracks were present in the sample stabilized with HALS-2 and a 20° gloss of 45 units (unpolished) was measured (Figure 4.75). This is a good indication that, even after six years, there are still enough nitroxyl radicals available to stabilize the coating.

As noted in Chapter 4.4.1, UV absorbers exhibit far better long-term stability when combined with HALS. Similarly, in considering the long-term stability of HALS, the effect of the UV absorber should not be neglected. As UV light is absorbed by the UV absorber, fewer radicals are formed, which, in turn, leads to fewer nitroxyl radicals and, therefore, a lower consumption of HALS.

As the vast body of results suggest, coatings can be stabilized for ten years or more, provided of course, that the right light stabilizer blends, consisting of HALS and UV absorbers, are used in the right concentrations.

5 Conclusions

Light is a wonderful thing. Without light, there would be no life on earth. All living organic species depend on light for their existence. We all love (sun) light, especially when we are on vacation. But light can also be harmful: extreme exposure to sunlight is dangerous for organic species, the human skin, plastics and organic coatings. Human skin, plastics and organic coatings can, however, also be protected from the harmful UV light with UV absorbers and free-radical scavengers.

In the coating field today, the principle of using a combination of UV absorbers and hindered amine light stabilizers (HALS) is the state of the art. A plethora of products is available on the market but there are significant differences in performance, ease of use, etc. Therefore, the criteria for stabilizer selection for a coating formulation are critical for a long service life of the coated article.

The short list of criteria for a product selection, as outlined in Sections 1 to 4 of this book, serve to answer the following basic questions:

- Application area (automotive, industrial, architectural, ...)?
- Targeted service life?
- Coating system (one-coat, two-coat, ...)?
- Binder systems and their light sensitivity per se?
- Type of substrate / need for non-migratory stabilizers?
- Spectral sensitivity of underlying substrate?
- Transmission requirements?
- Type of formulation (liquids containing solvents, solvent-free, powder)?
- Solvents used (organic and/or water)?
- Curing conditions (criteria: volatility, thermal permanence)?
- Film thickness (criteria: photo permanence, volatility of UV absorbers)?
- Other formulation ingredients (pigments, catalysts, other additives) and their possible negative interactions with the UV absorbers and/or HALS?

If a formulator considers these criteria when searching for the right light stabilizer package, he should be able to formulate a coating, which will survive dark side of this wonderful thing called "light".

Admoneri bonus gaudet,
pessimus quisque rectorem aperrime patitur. Seneca
(„A good man accepts reproof gladly; the worse a man is the more bitterly he resents it.")

6 References

[1] H. Hesse, from the poem "Nachts im April notiert" *(Written one night in April)*

[2] Römpp Chemielexikon Vol 3, Georg Thieme Verlag Stuttgart 1992

[3] A. Hofmann, J. Chem. Soc. 13 (1860), 87

[4] H. Bäckström, J. Am. Chem. Soc. 49 (1927), 1460

[5] H. Hock, W. Susemihl, Chem. Ber. 66 (1933), 61

[6] H. Bäckström, Z. Phys. Chem. Abt. B 25 (1934),122

[7] K. Päckel, J. Wagner, Arch. Pharm. 280 (1942), 373

[8] L. Meyer, W. Gearhart, Ind. Eng. Chem. 37 (1945), 232

[9] M. Neimann, E. Rozantsev, Y. Mamedova, Nature (London), 196 (1962), 472

[10] J. Rabek, "Mechanisms of Photochemical Processes and Photochemical Reactions in Polymers", John Wiley & Sons Ud., New York 1987

[11] J. Calvert, J. Pitts, "Photochemistry", John Wiley & Sons Ltd., New York, 1966

[12] B. Ranby, J. Rabek, "Photodegradation and Photostabilization of Polymers", John Wiley & Sons Ud., New York 1975

[13] R.T. Morrison, R.N. Boyd: Lehrbuch der Organischen Chemie; 3rd Edition, VCH, Weinheim 1986

[14] G. Korting, "Praxis der Dermatologie" (The practice of dermatology), Georg Thieme Verlag Stuttgart–New York 1982, 247 (Fig. 280)

[15] F. Liebel, S. Kaur, E. Ruvolo, N. Kollias, M.D. Southall; Journal of Investigative Dermatology (2012) 132, 1901–1907

[16] B. Dolezel, "Beständigkeit von Kunststoffen und Gummi" (The stability of plastics and rubber), Carl Hanser Verlag Wien 1978

[17] R. Wayne, "Photochemistry", Butterworth & Co., London 1970

[18] H. Becker, "Einführung in die Photochemie" (Introduction to photochemistry), Georg Thieme Verlag Stuttgart 1983

[19] F. Gugumus in H. Zweifel, R. Maier, M. Schiller (Ed) "Plastics Additives Handbook", 6th Ed, Carl Hanser Verlag Munich 2009, 139

[20] G. Berner, M. Rembold, Farbe & Lack 87 (1981), 930

[21] H.J. Heller, H.R. Blattmann, Pure and Applied Chemistry 30 (1972), 145

[22] G. Berner, G. Dexlex, E. Schoohf, FATIPEC Congress, Amsterdam 1980, Congress Handbook Vol. II, 61

[23] A. Valet, Farbe & Lack 96 (1990), 185

[24] P. Schirmann, M. Dexter, in "Handbook of Coating Additives", Marcel Decker Inc. New York/Basel1987, 225

[25] V.P.S. Judin, Chemistry in Britain, June 1993, 503

[26] H. Völz, G. Kämpf, G. Fritzky, Farbe und Lack 78 (1972), 1037

[27] W. Griebler, "Anorganische Weißpigmente" (Inorganic white pigments), Information (Information), Sachtleben Chemie GmbH, Duisburg 1987

[28] L. Dulog, U. Saupp, Farbe & Lack 99 (1993), 591

[29] J. McKellar, N. Allen, "Photochemistry of Man-Made Polymers", Applied Science Publishers Ltd, London 1979

[30] E.V. Schmid, in "Umwelteinflüsse auf Oberflächen" (Environmental influences on surfaces), Expert-Verlag Ehningen 1989, 81

[31] F. Gugumus in H. Zweifel, R. Maier, M. Schiller (Ed) "Plastics Additives Handbook", 6th Ed, Carl Hanser Verlag Munich 2009, 204

[32] F. Haselmeyer, Deutsche Farbenzeitung 31 (2) (1977), 52

[33] M. Rembold, H.E.A. Kramer, J. Soc. Dyers Colour. 94 (1978), 12

[34] T. Werner, H.E.A. Kramer, European Polymer Journal, Vol. 13 (1977), 501

[35] L. Wihan, Polymers Paint Colour Journal, Vol. 170, No. 4031 (1980), 696

[36] L. Avar, K. Bechtold, E. Hess, FATIPEC Congress, Budapest 1994, Congress Handbook Vol. III, 45

[37] A. Valet, Farbe & Lack 100 (1994), 919

[38] L. Avar, "2nd North American Conference on Stabilization and Degradation", Hilton Head (SC), 13–15 March 1995

[39] DEA 4 416 809 (Ciba-Geigy AG)

[40] USP 5.106.891 (Ciba-Geigy AG)

[41] J. Merril, R. Bennett, J. Chem. Phys. 13 (1965), 1410

[42] J. Pitts, H. Johnson, T. Kuwana, J. Phys. Chem. 66 (1962), 2456

[43] D. Williams, A. Heller, J. Phys. Chem. 74 (1970), 4473

[44] J. Otterstedt, J. Chem. Phys. 58 (1973), 5716

[45] G. Woessner, G. Goeller, P. Kollat, J. Stezowski, M. Hanser, U. Klein, H.E.A. Kramer, J. Phys. Chem. 88 (1984), 5544

[46] H.E.A. Kramer, Farbe & Lack 92 (1966), 919

[47] A. Weller, Z. Elektrochem. 60 (1956), 1144

[48] G. Woessner, G. Goeller, J. Rieker, H. Hoier, J. Stezowski, E. Daltrozzo, M. Neureiter, H.E.A. Kramer, J. Phys. Chem. 89 (1985), 3629

[49] EP 0180 548 (Ciba-Geigy AG)

[50] C. Decker, K. Zahouilhy, S. Biry, Polym. Degrad. and Stab. 47 (1995), 109

[51] J. Rieker, E. Lemmert-Schmitt, G. Goeller, M. Roesler, G. Stueber, H. Schettler, H.E.A. Kramer, J. Stezoswki, H. Hoier, S. Henkel, A. Schmidt, H. Port, M. Wiechmann, J. Rody, G. Rytz, M. Slongo, J. Birbaum, J. Phys. Chem. 96 (1992), 10225

[52] H. Shizuka, K. Matsui, Y. Hirata, I. Tanaka, J. Phys. Chem. 81 (1977), 2243

[53] H. Shizuka, M. Machii, Y. Higaki, M. Tanaka, I. Tanaka, J. Phys. Chern. 89 (1985), 320

[54] G. Stueber, M. Kieninger, H. Schettler, W. Busch, B. Goeller, J. Franke, H.E.A. Kramer, H. Hoier, S. Heukel, P. Fischer, H. Port, T. Hirsch, G. Rytz, J. Birbaum, J. Phys. Chem. 99 (1995), 10097

[55] M. Allan, T. Bally, E. Haselbach, L. Avar, Polym. Degrad. and Stab. 15 (1986), 311

[56] US 7'695'643 (Ciba Specialty Chemicals Corp.)

[57] J. Chaudet, J. Tamblyn, J. Appl. Poly. Sci. 8 (1964), 1949

[58] G. Scott, "Atmospheric Oxidation and Antioxidants", Elsevier, London 1965

[59] O. Cicchetti. Adv. Polym. Sci. 7 (1970), 70

[60] K. Schwarzenbach, B. Gilg, D. Müller, G.Knobloch, J-R.Pauquet, P. Rota-Graziosi, A. Schmitter, J. Zingg, E. Kramer in H. Zweifel, R. Maier, M. Schiller (Ed) "Plastics Additives Handbook", 6th Ed, Carl Hanser Verlag Munich 2009, 1

[61] A. Buchachenko, "Stable Radicals", Consultant Bureau, New York 1965

[62] B. Ranby, J. Rabek, "ESR-Spectroscopy in Polymer Research", Springer-Verlag Berlin, 1977

[63] V. Ivanov, V. Shlyapintokh, A. Shapiro, O. Khvostach, E. Rozantsev, Izv. Akad. Nauk SSSR, Sero Khim. 8 (1974), 1916

[64] D. Carlsson, D. Grattan, T. Suprunchuk, D. Wiles, J. Appl. Polym. Sci. 22 (1978), 2217

[65] E. Fremy, Ann. Chim. Phys. 15 (1845), 459

[66] O. Piloty, B. Schwerin, Ber. dtsch. Chem. Ges. 34 (1901), 1870

[67] K. Meyer, W. Reppe, Ber. dtsch. Chem. Ges. 54 (1921), 327

[68] H. Wieland, F. Kögel, Ber. dtsch. Chem. Ges. 55 (1922), 1798

[69] F. Banfield, J. Kenyon, J. Chem. Soc. 1956, 1612

[70] M. Rogers, J. Chem. Soc. 1956, 2784

[71] O. Lebedev, M. Khidekel, G. Razuvaev, Dokl. Akad. Nauk. SSSR 140 (1961), 1327

[72] E. Rozantsev, "Free Nitroxyl Radicals", Plenum Press New York-London 1970

[73] E. Denisov, Int. Symp. on Degradation and of Polymers, Brussels 1974, 137

[74] J. Pospisil, J. Pilar, S. Nespurek, J. of Vinyl & Additive Technology 2007, 119–132

[75] P. Klemchuk, M. Gande, E. Cordola, Polym. Degrad. and Stab. 27 (1990), 65

[76] E. Step, N. Turro, M. Gande, P. Klemchuk, Macromolecules Vol. 27, No. 9 (1994), 27

[77] E. Step, N. Turro, P. Klemchuk, M. Gande, Angew. Makr. Chem. 232 (1995), 65

[78] O. Kysel, P. Mach, M. Micov, Polym. Degrad. and Stab. 40 (1993), 31

[79] Y. Ohkatsu, J. Japan Petrol. Inst. Vol. 51, No.4, 2008, 191

[80] J.L. Hodgson, M.L. Coote, Macromolecules 2010, 43, 4573

[81] G. Gryn'Ova, K.U. Ingold, M.L. Coote, J. Am. Chem. Soc. 2012, 134, 12979

[82] K.U. Ingold, D.A. Pratt, Chem. Rev. 2014, 114, 9022–9046

[83] J. Gerlock, D. Bauer, D. Mielewski, Int. Conf. on Adv. in Stab. and Controlled Deg. of Polym., Lucerne 1989, 25

[84] D. Bauer, J. of Coat. Techn. Vol. 66, No. 835 (1994), 57

[85] D. Bauer, J. Gerlock, D. Mielewski, M. Paputa Peck, R. Carter, Polym. Degrad. and Stab. 28 (1990), 39

[86] D. Bauer, J. Gerlock, D. Mielewski, Polym. Degr. and Stab. 28 (1990), 115

[87] K. Bechthold, E. Hess, G. Lignier, Farbe & Lack 99 (1993), 25

[88] A. Marek, L. Kapralkova, O. Schmidt, J. Pfleger, J. Humlicek, J. Pospisil, J. Pilar, Polym. Degrad.Stab., 2006, Vol. 91, 444

[89] F. Gugumus in H. Zweifel, R. Maier, M. Schiller (Ed) "Plastics Additives Handbook", 6th Ed, Carl Hanser Verlag Munich 2009, 215

[90] N.J. Turro, Pure and Applied Chemistry 49 (1977), 405

[91] T. Förster, Z. Elektrochemie 54 (1950), 42–46 and 531–535

[92] A. Weller in "Progress of Reaction Kinetics", Pergamon Press London, Vol. I (1961), 187

[93] H. Heller, H. Blattmann, Pure and Applied Chemistry 36 (1973), 141

[94] D. Carlsson, D. Wiles, J. Polym. Sci. 12 (1974), 2217

[95] G. Cowell, Plastics Engineering 32 (1976) 10, 51

[96] F. Gugumus, Polym. Degrad. and Stab. 39 (1993), 117

[97] D. Mielewski, D. Bauer, J. Gerlock, Polym. Degrad. and Stab. 41 (1993), 323

[98] A. Jurgetz, H. Rothbächer, C. Bliefert, Farbe & Lack 91 (1985), 921

[99] A. Valet, F. Sitek, G. Berner, Farbe & Lack 94 (1988), 734

[100] H. Böhnke, E. Hess, Farbe & Lack 95 (1989), 715

[101] OICA – International Organization of Motor Vehicle Manufacturers, http://www.oica.net/category/production-statistics/2014-statistics/

[102] J. Ruf, "Organischer Metallschutz", Vincentz Verlag Hannover 1993

[103] L. Kahl, M. Bock, H. Blum, J. Pedain, XXII. Fatipec Congress Budapest 1994, Congress Handbook Vol. I, 270

[104] M. Bock, "Polyurethanes for Coatings", Vincentz Verlag Hannover 2001

[105] U. Meier-Westhues, "Polyurethane", Vincentz Network Hannover 2007

[106] W.J. Blank, Z.A. Heh, M. Picci, J. of Coat. Techn., Vol. 74, 926 (2002), 33

[107] A. Goldschmidt, H-J. Streitberger, "BASF Handbuch Lackiertechnik", Vincentz Network Hannover 2014

[108] M. Dornbusch, U. Christ, R. Rasing, "Epoxidharze", Vincentz Network Hannover 2015

[109] A. Noomen, XIII. Conf. Org. Coat. Sci. and Techn., Athens 1987, Congress Handbook, 251

[110] A. Noomen, Polymer Paints Colour Journal Vol. 176 (1986), 76

[111] M. Bock, R. Halpaap, Farbe & Lack 93 (1987), 264

[112] A. Valet, 1st Pacific Coating Forum, Shonan Village/Kanagawa (Japan), May 1995, Proceedings, 58

[113] G. Mathur, J. Kresta, K. Fritsch, Adv. Urethane Sci. Teehn. 6 (1978), 103

[114] R. Feiler, M. Curran, C. Bailie, Photodegradation and Photostabilization of Coatings, ACS Symposium Series Vol. 151 (1981), 183

[115] D. Bauer, Polym. Mol. Sci. Eng., 48 (1983), 737

[116] D. Bauer, R. Diekie, J. Koenig, Ind. Eng. Chem. Prod. Res. Dev. 25 (1986), 289

[117] D. Bauer, D. Mielewski, J. Gerlock, Polym. Degrad. and Stab., 38 (1992), 57

[118] D. Bauer, D. Mielewski, Polym. Degrad. and Stab. 40 (1993), 349

[119] A. Huber, Dissertation University of Stuttgart 1988

[120] U. Schutz, R. Schulze, J. Sickfeld, P. Trubiroha, XXII. Fatipec Congress Budapest 1994, Congress Handbook Vol. I, 218

[121] R. Lechler, Dissertation, University of Stuttgart 1992

[122] G. Berner, U. Kreibich, Adv. in Org. Coat. Sci. Techn. Ser. 4 (1982), 334

[123] L. Avar, R. Bill, E. Hess, Farbe & Lack 92 (1986), 915

[124] P. Schirmann, Techn. Pap. Soc. Manuf. Eng. (1990), FC90-675, 1

[125] G. Berner, M. Rembold, Farbe & Lack 89 (1983), 840

[126] K. Hoffmann, Farbe & Lack 76 (1970), 249

[127] ATLAS, "Sun Spots" Vol. 24, No. 47 (1994),1

[128] W. Heilen et al., "Additives for waterborne Coatings", Chapter 10, Light Stabilizers for waterborne coatings by A. Braig; Vincentz Nertwork Hannover 2009

[129] W. Peter, C. Schaller, C. Schellenberg, D. Rogez, "A new concept of aqueous additive preparations", PRA Conference Waterborne and High Solids Coatings, Brussels (2006)

[130] D. Bailey, O. Vogel, J. Macromol. Sci.-Rev. Macromol. Chem. 14, No. 2 (1976), 267

[131] P.V. Yaneff, K. Adamson, N. Cliff, M. Kanouni, JCT, Vol1, No. 3 (2004), 201

[132] N. Cliff, M. Kanouni, C. Peters, P.V. Yaneff, K. Adamson, JCT, Vol 2, No. 5 (2005), 371

[133] J. Knapczyk, J. Coat. Techn. 60 (1988), 63

[134] Anonym; Polymers Paint Colour Journal Vol.182 (1992), 126

[135] A. Valet, Polymer Paints Colour Journal Vol 185 (1995), 31

[136] USP 4.772.672 (Kansai Paint Ltd.)

[137] A.R. Marrion "The Chemistry and Physics of Coatings" 2nd Ed., The Royal Soc. of Chem. 2004, Athenaeum Press, Gateshead, UK

[138] W. Rutsch, K. Dietliker, D. Leppard, M. Köhler, L. Misev, U. Kolczak, G. Rist, XXth Int. Conf. in Org. Cont. Sci. and Techn., Athens 1994, Congress Handbook, 467

[139] B. Ranby, A. Hult, Org. Coat. 7 (1984), 137

[140] K. O'Hara, Polymers Paint Colour Journal Vol. 175, No. 4154 (1985), 776

[141] L. Gatechair, H. Evers, Modern Paint and Coatings 76 (1986), 38

[142] C. Decker, T. Bendaikha, Int. Conf. on Adv. in Stab. and Controlled Degr. of Polym., Lucerne 1985, Technomic, Lancaster 1989, 143

[143] C. Decker, K. Moussa, T. Bendaikha, J. Polym. Sci 29 (1991), 739

[144] A. Valet, D. Wostratzky, RadTech Report, November/December 1996, 18

[145] A. Valet, D. Rogez, Surface Coatings International Vol 6 (1999), 293

[146] W. Bäumer, Kontakte 3, Darmstadt 1989, 42

[147] A. Valet, Progress in Organic Coatings 35 (1999), 223

[148] T. Jung, A. Valet, G. Gscheidt, C. Decker, RadTech Asia, Yokohama (2003), conference proceedings, 742

[149] J. Baro, K. Dietliker, "Photoinitiators" in "Handbook of Industrial Inkjet Printing – A Full System Approach" (W. Zapka, Ed.), Wiley-VCH, Weinheim (2017 – in preparation)

[150] Clariant International Ltd., Muttenz, Switzerland, "Additves for Coatings", Edition July 2007

[151] J. Gerlock, A. Kucherov, R. Matheson, Polym. Degrad. Stab., 69 (2000), 1

[152] J. Gerlock, A. Kucherov, S. Smith, Polym. Degrad. Stab., 73 (2001), 201

[153] G. Berner, M. Rembold, Org. Coat. 6 (1984), 55

[154] A. Mar, Modern Paint and Coatings, October 1989, 162

[155] D. Bramer, M. Holt, A. Mar, Polym. Mat. Sci. Eng. 63 (1990), 647

[156] A. Valet, Farbe & Lack 96 (1990), 689

[157] M. Holt, D. Bramer, Int. Conf. on Adv. in Stab. and Controlled Degr. of Polym., Lucerne 1991, Congress Handbook, 23

[158] P. Brennan, C. Fedor, "Sunlight, UV & Accelerated Weathering", Soc. of Plastic Eng. Automotive RETEC (1987); Technical Bulletin LU-0822 (1994), The Q-Panel Company

[159] P. Brennan, G. Fedor, "Irradiance Control in Fluorescent UV-Exposure Testers", Technical Bulletin LU-8032 (1994), The Q-Panel Company

[160] G. Fedor, P. Brennan, "Comparison between Natural Weathering and Fluorescent UV-Exposures: UVA-340 Lamp Test Results", ASTM STP 1294, Am. Soc. for Testing and Material, Philadelphia, 1996

[161] J. Gerlock, M. Dean, T. Korniski, D. Bauer, Ind. Eng. Chem. Res. Dev. 25 (1986), 449

[162] L. Briggs, D. Bauer, R. Carter, Ind. Eng. Chem. Res. 26 (1987), 667

[163] J. Gerlock, D. Mielewski, D. Bauer, Polym. Degrad. and Stab. 26 (1989), 241

[164] P. Pappas, Progress in Org. Coat. 17 (1989), 107

[165] P. Gijsman, Polym. Prep. 34, 2 (1993), 183

[166] P. Gijsman, J. Hennekens, K. Janssen, Polym. Degrad. and Stab. 46 (1994), 63

[167] H. Hönel, Farbe & Lack 59 (1953), 179

[168] "Zwei Jahre Wasser-Basecoat in kritischer Betrachtung" (Two years waterborne basecoat looked at critically), Industrie-Lack "Schwerpunktthemen" (Subjects of special interests), Vol. 10, Vincentz Verlag Hannover 1989

[169] M. Gebauer, Praxis-Forum Tagung "Clearcoat-Konzepte in Entwicklung und Anwendung" Bad Nauheim 1996, conference handbook, 13

[170] D. Drake, Polymers Paint Colour Journal Vol 183, No. 4331 (1993), 318

[171] W. Kinza, Praxis-Forum Tagung "Die EPS-Praxis 1991", Bad Nauheim 1991, Conference Handbook, 48

[172] A. Uhl, Praxis-Forum Tagung "Die EPS-Praxis 1991", Bad Nauheim 1991, Conference Handbook, 76

[173] J. Thaler, Praxis-Forum Tagung "Clearcoat-Konzepte in Entwicklung und Anwendung", Bad Nauheim 1996. conference handbook, 48

[174] P. Clark, Automotive Body Painting, IBEC '94, Detroit, Michigan, USA, September 1994, 44

[175] F. Hartner, Praxis-Forum Tagung "Die EPS-Praxis 1995", Bad Nauheim 1995, Conference Handbook, 165

[176] D. Fink, G. Brindöpke, 3rd Nuremberg Congress, March 1995, Congress Handbook, Paper 29

[177] F. Witte, RadTech Europe 1995, Maastrich/NL, September 1995, Congress Handbook, 437

[178] "The advantages of plastics in Automotive", Preferred Plastics, Inc., January 31, 2014

[179] R. Ryntz, Adhesion to Plastics, Polymer Surfaces and Interfaces, M. Urban.(Ed.) Global Press, Chicago, Il., 1998

[180] F. Light., Mictrotome Technology – Distribution of Stabilizers in Multilayer Coatings and Plastic Coatings, Proc. of Addcoat 2001, Orlando, Fl., 2001

[181] G. Haake, F. Andrawes, B.H. Campbell, "Migration of Light Stabilizers in Acrylic/Melamine Clearcoats", Journal of Coatings Technology, 68, No. 855, 57 (1996)

[182] F. Andrawes, T. Valcaral, G. Haake, J. Brinen, "Depth Distribution of Light Stabilizers in Coatings analyzed by Supercritical Fluid Extraction-Gas-Chromatography and Time-of-Flight Secondary Mass Spectrometry", Anal. Chem. 70, p. 3762 (1998)

[183] Technical Information Makrolon®; Bayer Material Science, Edition 2004-09-28

[184] Exatec® Technologies, 2010

[185] R. Schwalm, "UV Coatings – Basics, Recent Developments and New Applications", Elsevier B.V. Amsterdam, 2007

[186] M.A. Masuelli, Materials Science "Composite Materials", January 23, 2013

[187] Technical Bulletin Jeffamine® Amines; Huntsman Corporation 2008

[188] T.J. Morgan, F.M. Kong, C.M. Walkup, Polymer 1984, 25, 375

[189] A. Braig, "Light Stabilization Tool Box – UV protection from UV to Near Visible"; PCI, March 2010

[190] A. Mar, P. Schirmann, "Stabilization of Maintenance Coatings», Waterborne and Higher Solids Coatings Symposium, New Orleans (USA), February 1987

[191] A. Valet, 1st Pacific Coating Forum, Shonan Village/Kanagawa (Japan), May 1995, Proceedings, 251

[192] K. Kronberger, D. Hammerton, K. Wood, M. Stödemann, JOCCA 11/1991, 405

[193] B. Perkett, Powder Coatings 8 (1985), 2

[194] D. Rogez, "Colour stabilization of wood and durability improvement of wood coatings: a new UV-light protection concept for indoor and outdoor applications", PCI magazine March, 2002

[195] J. Sell, W. Feist, Forest Products Journal Vol. 36, No.3 (1986), 57 and Nr.4 (1986), 37

[196] R. William, W. Feist, J. of Coat. Techn. Vol. 66, No. 828 (1994), 109

[197] W. Feist, Am. Paint. Coat., 27. March 1989

[198] F. Finus, Farbe & Lack 81 (1975), 604

[199] R. Sharrock, J. of Coat. Techn. Vol. 62, No. 789 (1990), 125

[200] K. Wagner, Hoechst "Kunstharz Nachrichten" 1986, 22

[201] H. Kastien, Farbe & Lack 95 (1989), 16

[202] D. Rogez, EUROCOAT Conference, Nice/France, September 1991, Congress Handbook, 348

[203] J. Heinskill, Farbe & Lack 100 (1994), 1011

[204] P. Gabriel, R. Lannucci, J. of Coat. Techn. Vol. 56, No. 712 (1984), 33

[205] G. Berner, M. Rembold, Org. Coat. 6 (1984), 55

[206] D. Bauer, J. Gerlock, R. Dickie, Prog. Org. Coat. 15 (1987), 209

[207] D. Bauer, J. Gerlock, R. Dickie, Prog. Org. Coat. 15 (1987), 209

[208] J. Pickett, J. Moore, Polym. Degrad. Stab. 42 (1993), 231

[209] C. Decker, K. Zahrouily, Polym. Mat. Sci. Eng. 68 (1993), 70

[210] C. Decker, Polym. Prep. 34 (1993), 139

[211] J. Bonekamp, L. Maeeker, J. Appl. Polym. Sci. 54 (1994), 1593

[212] J. Gerlock, W. Tang, M. Dearth, T. Korniski, Polym. Degrad. and Stab. 48 (1995), 121

[213] J. Catalan, F. Fabero, M. Guijarro, R. Claramunt, M. Santa Maria, M. de la Coneeption Foces-Foces, F. Cano, J. Eiguero, R. Sastre, J. Am. Chem. Soc. 112 (1990), 747

[214] M. Rembold, G. Berner, Org. Coat. and Appl, Polym. Sci. Proceedings 47 (1982), 624

[215] C. Decker, University of Mulhouse/ France, unpublished results

[216] J. Gerlock, T. Prater, S. Kaberline, J. de Vries, Polym. Degrad. and Stab. 47 (1995), 405

[217] J. Gerlock, T. Riley, D. Bauer, Poly. Degrad. and Stab. 14 (1986), 73

[218] D. Bauer, M. Dean, J. Gerlock, Ind. Eng. Chem. Res. 27 (1988), 65

Authors

Dr. Andreas Valet studied chemistry at Stuttgart University and received his PhD in 1985 under Prof. Dr. Dulog. He then joined the Additives Division of Ciba-Geigy AG in Basel, Switzerland (as of January 1997 Ciba Spezialty Chemicals, Inc. and as of 2009 part of BASF Switzerland AG) and began to work on light stabilizers for coatings. Three years later, he was appointed head of the light stabilizers development group. In 1995 he took charge of the "Radiation Curing" application group and from 1998 to 2001 he was responsible for the global R&D portfolio management of the Imaging & Coating Additives business unit. In 2001 he was named global head of R&D of Ciba's Coating Effects division and, later in 2004, global head of R&D for the Plastic Additive division. From 2009 on, he assumed various functions in the Basel Research Center of BASF Switzerland AG and is responsible today for innovation in the area of functional materials and interfaces. He has published numerous papers and patents on light stabilizers and photoinitiators and has given lectures at conferences and seminars both at home and abroad. He also authored the first edition of the present book in 1997.

Dr. Adalbert Braig studied chemistry at the University of Stuttgart, Germany and received his PhD in polymer chemistry in 1983 under Prof. Dr. L. Dulog. He then joined the former Additives Division of Ciba-Geigy AG, Basle, Switzerland (as of January 1997 Ciba Specialty Chemicals, Inc. and as of 2009 part of BASF Switzerland AG). He assumed various functions in the development of coatings additives both for Automotive and Industrial applications, initially focusing on corrosion inhibitors. 1990 to 1991 transfer to Ciba-Geigy Corp., Ardsley, NY (USA). As of 1998 main focus on the development of light stabilizers for coatings. In 2003 appointed to Senior Technical Fellow of Ciba Specialty Chemicals. He published numerous patents and papers on corrosion inhibitors and light stabilizers and gave lectures at conferences and seminars at home as well as the US and Japan. Starting May, 2011 he worked as Technical Manager Additives and Resins Europe until his retirement end 2015. He continues to support the development activities in the area of light stabilizers and his successor, Dr. André Kastler, as a consultant for the company.

Index

Symbols

2,2,6,6-tetramethylpiperidine 38
2K-PU 51
2P-PU 51
α-cleavage 67
α-hydroxyketones 67

A

absorbance 27, 63
absorption band 89
absorption edge 26, 55
absorption spectrum 25, 64, 67
accelerated weathering 79
acceptor 16, 22, 45
additive 11
air pollutants 49, 52
aminoether 43
antioxidant 36, 46
antioxidant, phenolic 36, 47, 55
antioxidant, primary 36
antioxidant, secondary 36
assessment criteria 84
automotive coatings 50, 51, 59, 80

B

basecoat 51, 53
basecoat stabilization 114
basicity 77, 126
benzophenones 24, 25, 26, 28, 33, 65, 90, 97, 135
benzotriazoles 24, 25, 26, 30, 65, 90, 97, 135
binder degradation 83, 119
binders 11
bis-acyl phosphine oxide 67
black box 79, 80
blistering 49

C

carbon black 22
catalysts 56, 65, 71
chain branching 16
chain cleavage 16
chain initiation 16
chromophore 21, 23, 36, 45, 63
clear coat 50, 51, 53, 55, 62, 64, 67, 69, 76, 83, 86, 90, 97, 103, 107, 113, 129, 135
coil coatings 122
colour change 49, 54, 83, 89, 111, 118, 126, 135
colour retention 86, 88, 94, 100, 104, 118, 135
compatibility 36, 56, 60, 71, 74
cracking 49, 54, 83, 84, 90
cyanoacrylates 24, 29

D

deactivation 16, 22, 28
degradation 11, 13, 16, 21, 22, 43, 45, 46, 51, 63, 85, 92, 111, 121, 123
Denisov cycle 42
dissociation energy 15
donor 16, 45

E

electron spin resonance (ESR) 36, 38, 137
energy conversion 24, 28
energy transfer 16, 22, 46
epoxy resins 52
extinction 27, 35, 46, 55, 117, 129
extraction 56, 71, 128, 137
extrusion temperatures 56, 57
exudation 56, 71, 74

F

filler 11, 51
film thickness 27, 55
fluorescence 22
fluorescent lamps 81, 82, 83
fluorescent radiation 45
Förster cycle 46
free radicals 16, 130
free-radical scavengers 21, 22, 36, 46

G

glass transition temperature 57, 92, 95, 101
ground state 15, 16, 22, 45, 128

H

HALS 38, 42, 52, 55, 71, 73, 74, 75, 76, 77, 90, 118, 131, 137
HALS concentration 44, 55, 119, 137
high-solids polyurethane coatings 51
high-solids TSA 51
hindered amine light stabilizers 38, 52
HS-2P-PU 51
HS-TSA 51
human skin 15
Hund's law 15
hydrogen bridge, inter-molecular 29
hydrogen bridge, intra-molecular 28, 29, 65, 128
hydroperoxides 17, 22, 46
hydroxybenzophenones 24
hydroxyphenyl-benzotriazoles 24
hydroxyphenylpyrimidines 24
hydroxyphenyl-s-triazines 24, 25, 26, 31, 32, 70, 97, 135

I

industrial coatings 79, 121, 123
interactions 65, 71
inter-molecular energy transfer 16
inter-system crossing 16, 27
intra-molecular energy transfer 16

J

Jablonski diagram 16

K

keto-enol tautomerism 28

L

Lambert-Beer's law 27, 55, 117
light 13, 15
light stabilizers 11, 22
long-term stability of HALS 137

M

metallic finishes 53
metal substrates 122
mode of action of HALS 42
mode of action of UV absorbers 27

N

nitroxyl radical 38, 42, 43, 77, 90, 137

O

one-coat finishes 51, 117
outdoor weathering 79, 80
oxanilides 24, 25, 26, 29, 30, 33, 90, 135

P

paint 11, 12, 16, 51
pendulum hardness 69, 78, 79
peroxide decomposing agents 21, 22, 36, 55
peroxide radicals 22, 43, 131
phenoxy radical 36, 37
phosphites 46, 55
phosphorescence 17, 27
phosphorescent radiation 45
photo-chemical process 15, 21
photo-chemical reaction 15, 23, 29, 53
photo-chemical stability 55, 128

photoinitiators 67, 103
pigment/binder ratio 118, 121
pigment degradation 86
pigments 11, 22, 51, 53, 114, 117, 126
piperidine nitrogen 44, 77, 79
pk_b value 77
polycarbonate 23, 109
polymer degradation 11, 14, 21, 22, 45, 52, 85, 86
powder coatings 52, 57, 100, 101, 102, 106, 125
primary requirements, UV absorber 55
proton transfer 28, 30

Q

Quencher 21, 22, 45

R

radiation 13, 17, 21, 70, 80, 85
radiation curing coatings 52, 56, 57
reactable HALS 75, 107
reactable UV absorber 61, 107
reinforced plastics 111
requirements, HALS 71

S

salicylic acid derivatives 24
secondary requirements, UV absorber 56
side reactions 64, 77
singlet oxygen 22, 29
singlet state 15, 22, 28
solubility 56, 60, 71
spin multiplicity 15, 16
stabilization options 21
stabilizers 11
sterically hindered amines 36, 38, 47, 52, 55, 58, 71
sterically hindered phenols 36
sterically hindered phenols, mode of action 37
storage stability 57, 101
sunlight 15, 55, 80

T

tetramethylpiperidine 77
thermogravimetry 74
thermosetting acrylics 51
thermosetting alkyd 52
titanium dioxide 22
transmittance spectrum 25
triplet state 15, 16, 29
TSA 51
two-coat 51, 53
two-pack polyurethane coatings 51

U

ultraviolet light 13
UV-A light 15
UV absorber 21, 22, 23, 25, 27, 30, 52, 55, 56, 58, 59, 61, 63, 64, 67, 117, 128
UV absorber concentration 55, 64, 117, 118, 135
UVA fluorescent lamps 81, 82, 83
UV-B light 15
UVB fluorescent lamps 81, 82
UV-curable clear coats 70, 103
UV-curable coatings 68
UV reflectance spectroscopy 129
UV spectroscopy 129

V

volatility 59, 71, 74, 128

W

weathering resistance 53, 55, 114, 122, 137
weathering tests 79, 81, 82
wood coatings 82, 125, 128
wood varnishes 106

Y

yellowing 35, 54, 64, 65, 67, 70, 76, 111, 126, 128
yellowness index 64, 65, 66, 70, 76, 84, 111, 126

www.ingramcontent.com/pod-product-compliance
Lightning Source LLC
LaVergne TN
LVHW021726021026
849161LV00004BA/738